Ox

n y

Pocket Atlas of Oxford

How to use this book

his book covers the central tourist and business disctrict, nd also the main suburbs of Oxford. Place names mentioned in the text are followed by a reference to their map position (e.g. Carfax Tower can be found on map **6** in the grid square **B3** i.e. **6B3**).

North is approximately at the top of the page.

Map 1 shows the whole city, its ring road, and surrounding villages.

The listings provided are by no means complete, and the selections are intended to suggest the range available rather than stand as a comprehensive guide. Every possible care was taken to ensure that all information given was accurate at the time of publication, but the publishers cannot accept responsibility for any errors causing expense, loss, or inconvenience.

The publishers would be grateful to learn of suggestions for improvements to this guide, and for any information which could be used to revise future editions.

The Legend, which appears on page 25 is crucial for accurate interpretation of the maps. Further notes on the maps are to be found on page 56.

Sightseeing

Oxford is known throughout the world as a centre of learn ing, but it is so much more than this. It is a place where a may find something in which to delight. Not only is the cit situated in the heart of England, it is a veritable jewel set i that heart.

Many of Oxford's ancient buildings are constructed fro local, Headington limestone which is a warm golden co our. Unfortunately it is rather soft and especially vulnerab to smoke-polluted air and rain. Its glory has howeve responded well to the results of the Clean Air Act and re cent careful restoration. Whilst this seems a never-endin task, it is undertaken with such expertise and dedicatio that the future seems assured.

The quality of light can always enhance or detract from view of an architectural subject. This is particularly true c Oxford. The situation and climate of the city, together wit the colour of the stone used mean that visually Oxfor responds well to sunlight, but especially to low sunligh Because of this, sight-seeing may be amply repaid betweer the months' of October to May.

Oxford has been the subject of many books through th years, and this slim volume cannot hope to compete in th depth, or the breadth of the whole range of them. It sets ou to provide the visitor and resident alike with pointers to th essence of the place, leaving the user free to read further a leisure on topics of personal interest.

Touristline, 24 hour service. Things to see and do locally 3 minute recorded message, frequently updated. Tel.24488

Walking in Oxford

The centre of Oxford is relatively flat, and it is therefor easy walking, even for those who are not physically fit. Th day tripper or visitor who wishes to make the best possibl use of their available time, might well take advantage of th guided tours organised by the **Tourist Information Centre** 6B3, in St. Aldates, for a moderate charge, Tel.726871 Footpaths are shown on the maps, but they are not ex haustive. Many of the colleges have worthwhile garder walks.

or the disabled: The lowering of many strategically placed kerbs has taken place. Cornmarket Street and Queen Street are both entirely kerb-less but buses, taxis, and emergency vehicles may use these otherwise pedestrian thoroughfares. The Covered Market and The Clarendon and Westgate Centres are generally on one level, or upper floors are serviced by lifts.

"Please could you tell me how to get to the University."

It is not really possible to locate The University precisely. There is no campus. Like Cambridge, Oxford University is still modelled on the medieval form.

The University co-ordinates such things as entrance qualifications, organisation, and the maintenance of academic standards i.e. degrees. Thus, whilst there are buildings all over the city which are university, rather than college buildings, the centres are not really physically unified. One might, however, well adopt the complex comprising the Bodleian Library, Sheldonian Theatre and the University Church as the ceremonial heart of the place. The Colleges are fiercly independent autonomous bodies which prepare students accepted by choice and merit to sit for degrees offered centrally by the University.

There are three terms in the University's academic year. Each term is of eight weeks' duration, and the weeks are numbered 1–8. The academic year starts in the Autumn with the **Michaelmas term,** the second, **Hilary term,** starts with the new year, and the third, **Trinity,** starts after Easter, in the late spring.

In 1878, Lady Margaret Hall, the first Oxford College for women was founded. Most colleges now accept both male and female students.

Places of Interest

All Souls' College: (Founded 1438), High Street, 6B4. Gatehouse – 1438; Chapel – 1442; Hall – 1730; Codrington Library – 1756; Front Quad – c.1440; Great Quad. – c.1725. Graduate College.

On the south wall of the Codrington Library is an ornate sundial designed by Christopher Wren.

Ashmolean Museum: (Founded 1683 – present site 1841 – 5). Beaumont Street, 6A2.
An Art Gallery as well as a Museum, named after Elias Ashmole. Its range is wide and it is one of the foremost museums in the country, as well as being the oldest.
Admission is free.

Balliol College: (Founded c.1263), Broad Street, 6A3.
Gatehouse – 1866; Chapel – 1856; Hall – 1877, Library – c.1430; Garden Quad. – 1714 until present; Front Quad. – 1867-9; Fisher Buildings – 1769; Salvin's Tower – 1853.
Founded by John Balliol as part of a penance resulting from a dispute with the Bishop of Durham.

Bodleian Library: (Opened 1602), Broad Street, 6B4.
Convocation House – 1634; Divinity School – 1427-90; Old School Quad. 1613-24.
The University Library is second only in scale to the British Library in London. It houses more than 4.9 million books. It is a copyright library, receiving a free copy of all books printed in England.
The Convocation House is where the ruling body of the University, Congregation, meets.
In the Divinity School the visitor finds more than just one of Europe's most beautiful rooms, for some of the Bodleian's treasures may be seen on display to the public.
The Old Schools Quad. is dominated by the Tower of the Five Orders (of architecture). The doors on to this quad. are inscribed with the names of the subjects which were formerly studied within. *See also New Bodleian Library.* Entry to the Divinity School exhibition is free.

Botanic Garden: (Founded 1621), High Street, 10A2.
Originally a 'Physick Garden' for the Faculty of Medicine, it has expanded beyond its original walls to include several large glass-houses, and a less formal garden area. The Daubeny Building was one of the earliest science laboratories in Oxford, but is now residential accommoda-

tion. Entry is free.

Brasenose College (B.N.C.): (Founded 1509), High Street, 6B4.
Gatehouse – 1509; Chapel – c.1660; Hall – 1509; Library – 1664; Frewin Hall (acquired 1580).
'Brasenose' refers to a door-knocker, (brazen nose), which pre-dates the foundation of the college.

Carfax. 6B3.
The cross-roads at the centre of the city. The origin of the name is obscure, but the Concise Oxford Dictionary gives it as a 'Place where (usually four) roads meet'.
Carfax Tower may be climbed for a small fee, and offers views over the city and especially down High Street. It is all that remains of the former City Church of St. Martin which was demolished in 1896 to widen the roadway. Over one hundred years before that, earlier road widening removed the Jacobean Carfax Conduit, now to be seen in newly-restored splendour at Nuneham Park, Nuneham Courtenay. On the tower is a clock with ornate quarter jacks which operate at the quarter hours.

Castle, New Road, 6B1.
Built by Robert d'Oilly in 1071, all that now remains of it, (within Oxford Prison), are the Mound, and the Tower and Crypt of St. George's Chapel. The chapel's dedication is believed to be among the earliest to England's patron saint (1074).

Cherwell, River. maps 4,7,10
Tributary of the Thames (Isis) joining the main stream south of Christ Church Meadow. It is not suitable for powered craft, but ideal for punting and canoeing. Punts may be hired from C. Howard and Son, at the Old Horse Ferry beneath Magdalen Bridge, and at the Cherwell Boat House, Bardwell Road.

Christ Church, Cathedral, (12th century onwards), St. Aldate's, 9A4.
The chapel of Christ Church College became Oxford's Cathedral when the See was moved from Osney Abbey in 1546, some years after the Dissolution of the Monastries.

The chapel had been the monastic church of S Frideswide's Priory. It is believed that had Wolsey com pleted his great concept it would have been pulled dow and replaced by a new chapel on the north side of th Great Quad. As it is, Wolsey had a section of the ol church removed to allow the new college to be built. Thi has left Oxford's Cathedral as one of the smallest in th country. It houses the shrine of St. Frideswide, a local sain Entrance fees are not charged to worshippers.

Christ Church, College, **(The House.)** Founded 152 (Wolsey) 1532 and 1546 (Henry VIII), St. Aldate's, 9A3.
Gatehouse – 1525 and 1681; Chapel (Cathedral) – 12th 19th century; Chapter House – 12th-13th century Cloister – c.1500; Hall and Kitchen – 1529; Great (Tom Quad. – North c.1650; remainder 1529; Peckwate Quad. – c.1710; Library – 1761; Canterbury Quad – c.1775; Meadow Buildings – 1865; Picture Gallery – 1968; Blue Boar Range – 1969; Memorial Garden – 1926.
This is the college of superlatives. Its common buildings ar generally larger and more spacious than those of other col leges, its Gatehouse must be one of the most striking an well-known buildings in Oxford. Tom Tower houses th bell 'Great Tom' which still rings out an obsolete curfew o 101 rings at 9.05p.m. each night.

City Walls.
Oxford was a walled town in medieval times. As time wen by the walls were breached from within as it expanded Only where special conditions have prevailed do the ol walls still survive.
New College's founder, William of Wykeham had t undertake to maintain them on his new site. His good fait is displayed by the longest stretch of surviving city wall—i New College Garden. The other considerable stretch run along Dead Man's Walk between Merton College Garden and Christ Church Meadows. Other small fragments still re- main but are not generally as available to the public a these.

Corpus Christi College; (Founded 1517), Merton Street, 9A4.
Gatehouse, Chapel, Hall, Library, First Quad. – 1512; Fellows' Building – c.1710; Thomas Building – 1928; Extension (Magpie Lane) 1884 and 1969.
In the First Quad. is a pillar bearing a sundial and an early 17th century perpetual calendar.

Exeter College: (Founded 1314), Turl Street, 6B3.
Gatehouse – 1701 and 1833; Chapel – 1859; Hall – 1618; Library – 1857; First Quad. – 17th-18th century; Palmer's Tower – 1432; Margary Quad. – 1965
The chapel is a 'copy' of the Sainte-Chapelle in Paris.

Folly Bridge, St. Aldate's, 9B3.
One of the sites proposed as the original Oxen-Ford, the river has been bridged at this point since the 13th century. The original bridge was the traditional site for Roger Bacon's astronomical observatory. 'Friar Bacon's Study' was a gate-tower spanning the roadway which later became known as 'The Folly'. Below Folly Bridge is the jetty where, in summer, Salter Bros. day and period pleasure boats may be hired. Their river steamers also start from Folly Bridge.

Green College: (Founded 1979), Woodstock Road, 3A1.
Gatehouse – 1979; Radcliffe Observatory – 1775. Graduate College.
Originally intended to be Radcliffe College, the name was changed to that of the benefactor whose gift made the college possible. It is a medical college with close ties with the Radcliffe Infirmary and other National Health Service facilities in the area.

Hertford College: (Founded 1874), Catte Street, 6B4.
Gatehouse – 1887; Chapel – 1908; Hall – 1889; Library – 1716; First Quad. – 17th century with additions; Second Quad. – Early 20th century; Holywell Quad. – 1976-81.
Hart Hall existed as far back as the 14th century. Eventually Magdalen Hall was granted the land and took over the site in 1822. Of interest to the visitor are the Chapel of Our

Lady at Smith Gate, (there was a gate in the City Wall o that name nearby), and the 'Bridge of Sighs' which crosse New College Lane near Catte Street.

History of Science, Museum: (opened 1935), Broad Street 6A4.
The Old Ashmolean Museum was opened to house Elia Ashmole's Tradescant bequest, and the first chemistr laboratory in England in 1683. The present museum house a collection of early mathematical, scientific, surgical an photographic instruments. Entry is free.

Holywell Music Room: (opened in 1742), Holywell Street 6A4.
Probably the earliest purpose-built concert hall in Europe It is part of the Music Faculty of the University.

Isis, River
Alternative, though now seldom used, name for the Rive Thames particularly associated with Oxford. It is thought t be a corruption of the Latin name of the Thame —'Tamesis'.

Jesus College: (Founded 1571), Turl Street, 6B3.
Gatehouse – 1571; Chapel – 1621; Hall – 1617; Library – 1679; First Quad. – c.1600; Second Quad. – c.1680; Shi Street Buildings – 1905; Old Members' Building – 1971 Turl Street Front – 1756 and 1865; Stevens Close – 1975

Keble College: (Founded 1870), Parks Road, 3B3.
Liddon and Pusey Quads. – 1870; Hayward and De Breyn Quads. – 1970.
The college was conceived and constructed in two stages the original at the time of the foundation, in red brick, ful of Victorian flamboyance and confidence; and the second in celebration of its centenary in massive twentieth centur style. Whilst not a theological college, Keble ha considerable links with the Church of England.

Keble Triangle, Banbury Road, 6A2.
An assemblage of largely 20th century University Buildings most prominent of which are the Engineering Science

Block, the fan-shaped Nuclear Physics Block, and the Department of Metallurgy.

Lady Margaret Hall (L.M.H.): (Founded 1878), Norham Gardens, 11B3.
First of the Oxford women's colleges, its main entrance is from Norham Gardens. The college has beautiful grounds which include riverside walks.

Linacre College: (Founded 1962), St. Cross Road, 4B2.
Graduate College occupying an early 20th century building which originally housed a Roman Catholic convent.

Lincoln College: (Founded 1427), Turl Street, 6B4.
Front Quad. – c.1450; Hall reconstructed – 1791; Kitchen – c.1437; Chapel – c.1630; Library (formerly All Saints' Church) – 1706-8; Chapel Quad. – 1608-31; Grove Building – 1880.
Despite its later reconstruction the Hall, dating from 1437, still has the octagonal louvre in the roof which allowed smoke to leave the room prior to the installation of a fireplace in 1699. They were once common, but now this is the only one left in an Oxford college hall.
Off the Front Quad. John Wesley's room may be seen.

Magdalen College: (Pronounced 'Maudlin'), (Founded 1458), High Street, 7B3.
Gatehouse – 1884; St. John's Hospital – 13th century; Kitchen – 13th century; Chapel – 1475; Hall – 1475; Cloister Quad. – 1475; Founder's Tower – 1485; Muniment Tower – 1488; Bell Tower – c.1500; Grammar Hall – 1614; New Buildings – 1733; St. Swithun's Quad. – 1880; Longwall Quad. – 1931; Waynflete Building – 1963.
Magdalen College is possibly best known for two things, first for its Bell Tower and second, for its choir. The two come together each May Morning when as a prelude to the other celebrations the chapel choir sings an old Latin hymn from the top of the tower at 5.00 a.m. G.M.T.
There are deer in the park, and a delightful waterside footpath round Addison's Walk.

Martyr's Memorial: (Built 1841), St. Giles' Street, 6A2.
Commemorates the martyrdom of the Anglican Bishops Cranmer, Latimer, and Ridley, who were burned at the stake in Broad Street during the reign of Queen Mary Tudor, (1555-6).

Merton College: (Founded 1264), Merton Street, 10A1.
Gatehouse – 1418; Chapel – 1290-1424; Hall – 1277; Library – 1373-8; Mob Quad. – 1304-11; Fellows' Quad. – 1610; Grove Buildings – 1864 and 1930; St. Alban's Quad. – 1905-10.
Probably the first true college in the university, its apparently haphazard layout of buildings formed the pattern for subsequent colleges to follow. Scholars are sometimes known as Postmasters, (Postmaster is a corruption of the Latin 'portionistae'). Postmasters Hall is on the site of some of the earliest buildings. The college also boasts a court for the ancient game of Real or Royal Tennis (1595). Mob Quad. is the oldest quadrangle in Oxford, and the Library is the oldest in England still in its original form. The Chapel is the largest of the Oxford college chapels if the Cathedral is discounted. Around Merton gardens is a considerable portion of the old City Wall, but this is not open to the public except from the outside along Dead Man's Walk.

Mesopotamia, The, 4B3.
Name given to the island between the River Cherwell's braids from the University parks, to Magdalen College Sports Ground. The island carries a very pleasant waterside walk.

Museum of Oxford, St. Aldate's, 6B3.
As the name suggests, this museum, sited next to the Town Hall on the Blue Boar Street corner of St. Aldate's, details the growth of both City and University (Town and Gown). It has a book shop and many interesting exhibits.

New Bodleian Library, (Built 1940), Parks Road, 6A4.
Joined to the older library buildings by underground conveyor, the New Library is not generally open to the public.

New College: (Founded 1379), New College Lane, 7B1.

Gatehouse – 1380; Chapel, Hall, Library and First Quadrangle – 1386; Garden Quad. – c1700; Cloister and Bell Tower – 1400; Robinson Tower – 1868; Holywell Buildings – 1896; Memorial Library – 1939; Sacher Building – 1962; City Walls–13th and 14th century.
New College was a grand concept from the start. No expense was spared and the result is magnificent. The buildings are matched by the gardens which are enclosed on the north and east sides by the longest, and best preserved section of the City Walls.

Nuffield College: (Founded 1937), New Road, 6B1.
The buildings date from 1955 – 60. Graduate college.
The college was founded by Oxford's most illustrious industrialist/philanthropist, William R. Morris—Lord Nuffield. The name Nuffield is associated throughout Oxford (and indeed the world) with the motor industry and philanthropic donation.

Oriel College: (Founded 1326), Oriel Square, 6B4.
Gatehouse – 1620; Chapel, Hall and First Quad. – 1642; Library – 1788; Second Quad. – 1720; St. Mary's Quad. – 1640; Rhodes Building – 1911.
An Oriel is a type of projecting window supported on corbels, (a fine example may be seen in the Balliol Master's Lodgings). Alternatively it has been suggested that the name comes from that of a building 'La Oriole' presented by King Edward II for the new college. St. Mary's Quad. contains what remains of St. Mary's Hall. The south-west corner dates from the middle of the 15th century.

Oxfam House: Banbury Road, 11A2.
Headquarters of the Oxford Committee for Famine Relief (Oxfam) which co-ordinates relief work throughout the world.

Oxford Union Society: (Founded 1823), St. Michael's Street, 6B2.
Originally a Literary, Debating and Social Club for students. It has won international renown for its debating, and has been a proving ground for many a future senior politician or M.P., indeed many are life members. The Debating Hall itself dates from 1878. There is a considerable Library. It is

nothing to do with the **Oxford University Union of Students**, who have an office in Little Clarendon Street, 3B1.

Painted Room, Cornmarket Street, 6B3.
This second floor room is traditionally that in which William Shakespeare slept on his journeys from Stratford-upon-Avon to London. In Shakespeare's day the building was the Crown Inn. It looks an 18th century structure but this is only a façade, and behind it is the 15th century inn. The paintings on the walls were discovered in 1927 and are well preserved. Entrance is by appointment only.

Pembroke College: (Founded 1624), Pembroke Square, 9A3.
Gatehouse – 1673-94 and 1830; Chapel – 1732; Hall – 1846; First Quad. – 1626-70 and 1830-40; Second Quad. – 1844-48; North Quad. – 16th century to the present day; Besse Building – 1962; McGowin Library – 1975; Macmillan Building – 1977.
Founded on the site of the medieval Broadgates Hall and incorporating Cardinal Wolsey's Almshouses (the Master's Lodgings) dating from the early 16th century, Pembroke's southern boundary marks the line of the old City Walls. Its name derives from William, Earl of Pembroke.

Pitt Rivers Museum, (Founded 1885), Parks Road, 3B4 & 11B2. Founded by the gift of a collection of artifacts from General A. H. Lane Fox Pit Rivers this Museum of Ethnology and Pre-History has so much material for display that it is in the process of dividing itself between two sites. The original location is at the back of the University Museum – through which entry is gained to it. (Its times of opening are rather more restricted than those of the University Museum, but entry is free.) The new location is along the Banbury Road.

The Queen's College: (Founded 1341), High Street, 7B1.
Gatehouse and High Street frontage – 1735; Chapel – 1714-19; Hall – 1715; Library – 1695; First Quad. – 18th century; Second Quad. – c.1700; New Quad., (opposite side of Queen's Lane, between St. Ed-

mund Hall and High Street) – 1970; Florey Building – 1971.
Founded by Robert de Eglesfield under the patronage of Edward III's Queen Philippa the college catered largely for students from the north of England. Because of the difficulty of travel in those far off days many members stayed in the college over the Christmas and New Year period. This led to the growth of two picturesque customs: the Boar's Head celebration (and its associated carol) and the Needle and Thread Feast on New Year's Day.
The Queen's College was virtually refounded in the 17th and 18th century when the cost of much new construction was contributed by Queen Caroline, (1733). Thus it is Queen Caroline who looks down from the gatehouse cupola.

Radcliffe Camera, (Built 1748), Radcliffe Square, 6B4.
Dr. John Radcliffe left a bequest for the foundation of a library which has since become one of the most famous buildings in the city. Designed by James Gibbs it was originally a science library but is now part of the Bodleian complex. It is used as a general reading room and is not open to the public. The name camera means 'vault', or refers to having an arched cover in this context.

St. Anne's College: (Founded 1879/1952), Woodstock Road, 3A1.
Gatehouse – 1966; Hall – 1959; Library (Hartland House) – 1936; Rayne Building – 1968; Wolfson Building – 1964.
Womens College founded originally as the Society of Oxford Home Students.

St. Antony's College: (Founded 1948), Woodstock Road, 11B2.
Financed by the gift of M. Antonin Besse the college took over the buildings of an anglican women's community which had been built in 1868. The old buildings are now mainly used as a library, the new building in the middle of the gardens contains the Hall, and other services. St.Anthony's is a graduate college. The entrance is on the Woodstock Road.

St. Catherine's College (St. Cat's): (Founded 1868/1962), Manor Road, 7A4.
The buildings are among the most striking modern buildings in the university. The whole conception was the work of one man, Arne Jacobsen, right down to the cutlery used in hall. The complex was turned over to its students finally in 1977. The chapel is the ancient Norman church of St. Cross, 7A2.

St. Cross College: (Founded 1965), St. Giles' Street, 6A2. Graduate College which has recently moved into the Pusey House premises in St. Giles', but it maintains an annex in St. Cross Road.

St. Edmund Hall (Teddy Hall): (Founded before 1270), Queen's Lane, 7B2.
Gatehouse – c.1741; Hall – c.1660; Chapel – 1680; Library (formerly St. Peter-in-the East church – 12th-13th century); South Range – 1934; Besse Building – 1969.
Named after St. Edmund of Abingdon whose monastic cell was on the north side of the church of St. Peter-in-the East. It is a good example of the concept of a medieval Hall, although its architecture is much later.

St. Hilda's College: (Founded 1893), Cowley Place, 10B3. On the site of Milham Ford and occupying buildings built originally in 1775 it has had several additions in the 20th century.

St. Hugh's College: (Founded 1886), St. Margaret's Road, 11B2.
Named after the 13th century Bishop of Lincoln in whose See Oxford was, St. Hugh's was started for women students who could not afford the fees at Lady Margaret Hall. It has been on its present site since 1916. The main entrance is in St. Margaret's Road, and its buildings are all modern.

St. John's College: (Founded 1555), St. Giles' Street, 6A2. Gatehouse and First Quad. – 1437; Chapel – 1530; Old Library – 1596; Laud's Library – 1636; Canterbury Quad. – 1636; North Quad. – 1880-1959; Dolphin

Quad. – 1948; Sir Thomas White Building – 1975.
Founded as St. Bernard's College, and later developed into one of Oxford's most splendid colleges by Archbishop Laud, it also boasts beautiful gardens.

St. Mary the Virgin Church, (Built 1189 onwards), High Street, 6B4.
One of the largest and finest churches in Oxford, St. Mary the Virgin is the University Church, where the University Sermons are preached. It has been used in the past for many University Ceremonies, and even trials.
Of special note is the fine English Baroque south door which has just been restored. Also on the site are the Old Congregation House and a Brass Rubbing Centre. The tower may be climbed for a small fee, and offers interesting views over the college roof-tops and down High Street.

St. Peter's College: (Founded 1928), New Inn Hall Street, 6B2.
Gatehouse and Library, (Linton House) – 1797; Hannington Hall – 1833; Chapel (St. Peter-le-Bailey Church) – 1874; other buildings are of more recent origin, except for the Master's Lodging (Canal House) – 1828.
As a Hall, St. Peter's has been in existence since the mid-nineteenth century. The old rectory building (now Linton House),and the church, then newly moved and somewhat redesigned from its former Queen Street site, became the nucleus of the new foundation.

Sheldonian Theatre, (Built 1669), Broad Street, 6A4.
Designed by Christopher Wren and based on the Theatre of Marcellus in Rome it is where Encaenias (degree ceremonies) are held at the end of the summer term. The Lantern offers views over the heart of the university. There is a small charge for entry.

Somerville College: (Founded 1879), Woodstock Road, 3B1.
Gatehouse and Eastern Quad. – 1934; Chapel – 1935; Dining Hall (Maitland Building) – 1913; Library – 1904; Walton House – 1828; Fry Building – 1964; Wolfson Building – 1967.

Womens college centred round the old Walton Manor House. Somerville is one of the oldest womens colleges.

Trinity College: (Founded 1555), Broad Street, 6A3.
Gatehouse (Trinity Cottages) – c.1690; Chapel – 1694 Hall – 1618; Old Library – 1417; Chapel Quad. – 1421 Garden Quad. – 1682; Kettell Hall – c.1620; New Buildings – 1885; New Library – 1928; Dolphin Gate – 1948; Cumberbatch Quad. – 1968.
Originally called Durham College prior to the Reformation Trinity is without an imposing gatehouse. The Chapel Quad. is all that remains of Durham College, and it is also known as Durham Quad. There is a pleasant garden containing an ancient Lime Walk.

University College (Univ.): (Founded 1249), High Street 7B1.
Gatehouse – 1638; Chapel – 1666; Hall – 1656 and 1904; Library – 1861: First Quad. – 1675; Radcliffe Quad. – 1719, Shelley Memorial – 1894; Goodhart Quad. – 1962.
This college, whilst not founded by King Alfred as was at one time thought, is probably the oldest Oxford college Little now remains of the medieval buildings and in general the surviving structure is of 17th century or later date. On the outside wall a plaque commemorates Robert Boyle (Boyle's Law), whilst inside there is a memorial to the poet Shelley.

University Museum, (Built 1855), Parks Road, 3B3.
Amazing Victorian building which is an education in itself. This lofty, red-brick, glass-roofed, many pillared cathedral of natural history is worth a visit for the buildings' diversity alone, but its contents are also impressive and well displayed. The grass court in front of the building covers the Radcliffe Science Library. When this was being excavated, Roman burials were discovered, which rewrote Oxford's early history. It had been accepted previously that the Romans by-passed the marshy, unhealthy area that has become the centre of the city. On passing through the University Museum the visitor gains access to another wonderful storehouse of antiquities, the Pitt-Rivers Museum of Ethnology and Prehistory.

niversity Press, (Built 1829), Walton Street, 2B4.
1478 Oxford University licensed Theodoric Rood to rint a commentary on the Apostle's Creed. In 1978 the)xford University Press celebrated the Quincentenary of rinting in Oxford. O.U.P. is a printing and publishing ouse of international repute and proportions. It is, erhaps, best known for the publication of Bibles, Prayer ooks, and Dictionaries.

Iniversity Science Area, South Parks Road, 3B4.
n area or rather areas devoted to the study of science and echnology. The main area is enclosed in the angle made y South Parks Road and Parks Road; (recent development outh of South Parks Road has extended this). The secon-ary area is that enclosed by Parks Road, Banbury Road, nd Keble Road and called the Keble triangle. This too has xtended itself across the Banbury Road. The other area pans the Banbury Road between Bevington Road and Norham Road.

Vadham College: (Founded 1610), Parks Road, 6A4.
All the older buildings are 17th century; The Holywell pro-perties are 17th and 18th century; New Building – 1952; New Library – 1977.
Built on the site of a monastic foundation just outside the ld city's Smith Gate, the college is a fine example of 17th entury architecture and has beautiful gardens.

Wolfson College: (Founded 1966), Linton Road, 11A3.
Modern graduate college which took up its present buildings in 1974. Its original concept was as a mixed foun-dation where men and women graduates could continue heir studies for higher qualifications. The main gate is at he bottom of Linton Road, and the college is sited in the Cherwell water meadows.

Worcester College: (Founded 1714), Walton Street, 6A1.
Gatehouse and Library – c.1791; Hall – 1784; Library – c.1730; Pump Quad. and South Range – 15th cen-tury; North Range – 1776; Nuffield Block – 1938; Besse Building – 1954; New Building – 1961; Wolfson Building – 1971; Sainsbury Building – 1982.

In 1298 Gloucester College was founded on the preser
Worcester site, for monks from Gloucester. After th
Dissolution of the Monasteries the buildings were repaire
and became known as Gloucester Hall. The quaint co
tages, (camerae), which form the South Range are relics c
monastic days, as is much of Pump Quad. There are quit
extensive gardens, and a large lake.

Parks and gardens.

By its nature Oxford abounds in green areas. Most college
have formal gardens worth visiting, but some have mor
extensive informal or 'natural' areas, such as **Chris
Church,** 9A4, **Magdalen,** 7B3, and **Worcester,** 5A4. Th
green on the maps is not intended to show areas to whic
the public generally have access, but to suggest areas c
green open space including wooded land. A large amour
of this is taken up by college playing-fields. The footpath
shown are selected, and do not represent any public righ
of way. Only the longer college walks are included.
Colleges are private property, but the general public ar
allowed free access to parts of tourist interest at regula
times during the day. These times are prominentl
displayed in the gatehouse or lodge, and are in some case
further restricted on Sundays and in 'Term' when th
undergrauates are at their studies.
In the context of the University, mention should be mad
here of the **University Parks,** maps 3 and 4, and th
Botanic Gardens, 10A2. Both are well worth visiting.
Most Local Authority parks have a variety of recreationa
and sports facilities which are open for public use on pay
ment of a fee. The facilites vary from park to park, but in
clude Putting, Bowls, Field Hockey, Cricket, Soccer, Hurl
ing, Rugby, Angling, Boating, Open-Air Swimming, an
Tennis. In addition there are children's swings, slides etc. in
most and paddling pools in some.
Many parks offer several of these activities, but those in
volving team use are usually booked regularly in advanc
by local clubs.
There are many recreation grounds around the city.

ort Meadow (with Wolvercote Green), 1B2, is common ınd. It is neither park nor private. It is part of the flood lain of the River Thames, and is consequently pretty flat. It allowed to flood in the winter to relieve more vulnerable esidential areas down-stream. As a side affect this enor-ıous area of shallow water freezes in a hard winter offer-ıg free skating to all, in addition to its other recreational ıcilities.

hese include angling, hour, day and period hire boats and ruisers, sailing, swimming and model aircraft flying. The ver towing path offers easy walking and delightful sur-oundings. There is parking, and access from Walton Well oad, 2A2, and Godstow Road, Wolvercote, 1B1. There re further access points from Aristotle Lane, 11B1, Binsey ane near the Perch Inn, 1C2, Godstow Road near the ailway bridge, 1B2, and by water-side walk from the otley Road at Osney Bridge, 5B1, or the Oxford Canal, A3.

hotover Hill Country Park offers a variety of outdoor ctivities including miles of nature walks, through wooded, crub, and open areas. Access is from Old Road, leadington, Wheatley, or Horspath.

River Use

ıll craft used on the waterways around Oxford must be egistered, and motorised boats require a licence. These re available from the Thames Water Authority, or The ritish Waterways Board (Canal and Parts of Cherwell)

ngling in the area requires the minimum of a T.W.A. Rod icence, and usually a more specific permit as well. These re available from local fishing tackle shops.

here are several river bathing places which are open from Aay to September only:

Godstow Road, 1B1 (Thames).

Long Bridges Bathing Place, Donnington Bridge Road, A2 (Thames).

Tumbling Bay Bathing Place, Botley Road (access), 5A1 Thames).

Parson's Pleasure, South Parks Road (access), Men only, Cherwell), 4A3.

View Points

Often view points (shown in red on the maps thus -•-) are thought of as being places where the subject to be viewed may be looked down upon. There are such places, but several of the locations listed below are from road or footpath and at similar elevation to the city. Thus it is suggested that the disabled may also enjoy its glories without difficulty.

Town

Carfax Tower, 6B3—a small charge is made, and there are steps to reach the open top.
St. Mary the Virgin (University) Church, 6B4—a small charge is made, and there are steps to reach the open top.
Sheldonian Theatre Lantern, 6A4—a small charge is made and there are steps to reach the enclosed top.
Crescent Road, Temple Cowley, 15A3—the top of the hill.
High Street, maps 6B, 7B—proceed from end to end in either direction, (preferably both), and watch as one of Europe's finest communities of architecture unfolds before you.
Lenthall Road, Rose Hill, 1D3—open road view over allotment gardens.
Merton Street, 9A4-7B2—the east-west section is a cobbled street with some of the earliest collegiate buildings, and is full of character.
Radcliffe Square, 6B4—the view from any of the four entrances is magnificent.
South Park, 13B2—the eastern (top) end.

Around

Boar's Hill, 1D1—one of the classic views of the city. Jarn Mound gives general views of Oxford and the surrounding countryside.
Elsfield, 1B3—general views of the city.
Godstow Lock, 1B1—the towing path of the River Thames offers an unique view of the spires and towers of the city, and easy walking as well.

Railway approach from the south, 1D2—a moving panorama of the architectural glories of Oxford, especially fine in afternoon sunshine.
Raleigh Park, Hinksey, 1D2—general views of the city.
Shotover Hill, Headington, 1C3—a country park offering general views of the Cowley car assembly plants and Blackbird Leys Estate.
Southern By-Pass Ring Road, 1C1—this section, between the Cumnor, and Abingdon By-Pass interchanges, (named before the Ring Road was completed), offers several views of the city. The view from its southern end approaches that which gave birth to the 'City of Dreaming Spires' description.
Wolvercote Green,1B2—the northern extension of Port Meadow. Access with car parking at the Bathing place on the Godstow Road.

Entertainment

Theatres

Apollo Theatre, George Street, Oxford, 6B2.
Pegasus (Youth) Theatre, Magdalen Road, Oxford, 14A4.
The Playhouse (incl. The Burton Rooms), Beaumont Street, Oxford, 6A2.

Cinemas

ABC Studio 1,
ABC Studio 2, George Street, 6B1.
ABC Studio 3,
ABC Super, Magdalen Street, 6A2.
Not the Moulin Rouge, New High Street,13A4.
Penultimate Picture Palace (Club), Jeune Street.
Phoenix Studio 1
Phoenix Studio 2 Walton Street, 2A4.

Museums and Art Galleries

Ashmolean Museum, Beaumont Street, 6A2. Also Art Gallery. *(See places of interest section).*

Bodleian Library, Divinity School, Broad Street. *(See places of interest section).*

Christ Church Picture Gallery, Canterbury Quad., 9A4.

Museum of the History of Science, Broad Street, 6A4. *(See places of interest section).*

Museum of Modern Art, Pembroke Street, 9A2.

Museum of Oxford, St. Aldate's, 6B3. *(See places of interest section).*

Open-air Art Exhibitions, on fine sunday afternoons from Spring—Autumn, exhibits hang on the Parks Road railings of the University Parks, 3A3.

Pitt-Rivers Museum of Ethnology, Parks Road, 3B4. Pitt-Rivers Museum of Ethnology (extension), Banbury Road, 11B2. *(See places of interest section).*

Rotunda (Doll) Museum, Grove House, Iffley Turn, 14B4.

University Museum, Parks Road, 3B3. *(See places of interest section).*

There are also some commercial Art Galleries in the city, and numerous specialist antiquarian shops. Being Oxford, browsing is encouraged in such places.

Main Concert Venues

Holywell Music Room, Holywell Street, Oxford, 6A4.

Maison Française, Norham Road, Oxford, 11B2.

Old Fire Station Arts Centre, George Street, Oxford, 6B1.

Sheldonian Theatre, Broad Street, Oxford, 6A4.

St. Paul's Arts Centre, Walton Street, Oxford, 2B4.

Town Hall, St. Aldate's, Oxford, 6B3.

Music at Oxford, uses mainly College Venues.

Other Performing Arts Venues

Christ Church Cathedral, St. Aldate's, Oxford, 9A4.

Clarendon Press Centre, Walton Street, Oxford, 6A1.

Newman Rooms, Roman Catholic Chaplaincy, St. Aldate's, Oxford, 9A3.

Churches and Church Halls—various

College sites—various

Other forms of Entertainment

Association Football is played professionally on the Manor Road Ground, London Road, Headington, 13A4. **Oxford United F.C.** play in the First Division of the Football League.

Oxford's semi-professional football team is **Oxford City F.C.** and plays on the Whitehouse Ground, Abingdon Road, 14A1.

Athletics meetings are held on the University Running Ground, Iffley Road, 14A3.

Boating is available on the River Thames, and to a more limited extent on the River Cherwell. The Cherwell is shallow and unsuitable for cruisers. It is, however, the Oxford home of the punt.

Cherwell Boat House, Bardwell Road, 11B3 (Cherwell).
Howard and Son, Old Horse Ferry, Magdalen Bridge, 7B3 (Cherwell).
Medley Boat Station, Port Meadow, Walton Well Road, 1C2 (Thames/Canal).
Salter Bros. Ltd., Folly Bridge, St. Aldate's, 9B3 (Thames).

Circuses come regularly to Oxford and are sited in the Oxpens area, 8A4.

First-class **County Cricket** is played in the University Parks, 3A4, by Oxford University Cricket Club.

Firework Display, held on, or conveniently near Guy Fawkes day (November 5th) in South Park, 13B2. It is organised by the Oxford Round Table.

The city **Golf Courses** are: Southfield Golf Club, Hill Top Road, 13B3 (Club House), 15A2 (Course), and North Oxford Golf Club, Banbury Road, 1B2.

Greyhound Racing is staged at Oxford Stadium, 1D3.

Ice Rink popular new addition to the city's list of attractions (opened 1984), 8A4.

Lord Mayor's Parade progresses from the Woodstock Road, outside the Radcliffe Infirmary, 3A1, via Carfax to South Park, 13B2, where celebrations continue until the end of the day. It takes place on the Late Spring Bank Holiday.

May Morning celebrations traditionally start with the singing of a Latin hymn from the top of Magdalen College Tower, 7B3. Then, led by Morris Dancers, the assembled multitude proceed up High Street to Radcliffe Square, and the New Bodleian Library Building, where performances are given. Other celebrations of an informal and often impromptu nature go on for much of the day.
Rowing events are held on the River Thames (Isis) between Iffley Lock, 14B3, and the College Boathouses, 14A2. These include the Torpids (Hilary Term), and the Eights Week in Trinity Term.
Rugby Union Football is played on the Oxford University Rugby Ground, Iffley Road, 14A3.
St. Giles' Fair is held in St. Giles' Street, 3B2, on the first monday and tuesday after the first sunday after the 1st September (St. Giles' day), each year. **Smaller Fairs** are sometimes found in Cutteslowe Park, 1B2, or on Wolvercote Green, 1B2.
The Sheriff's Races since 1980 these are run on the old Oxford Race Course site, Wolvercote Green, 1B2. They have become an annual event. The horse races are not held under Jockey Club Rules.
Speedway (motorcycle) is staged at Oxford Stadium, 1D3. The Team is the Oxford Cheetahs.
Stock Car Racing is staged at Oxford Stadium, 1D3.
Indoor Swimming baths open to the public are, Temple Cowley Baths, Temple Road, 15B3, and the Ferry Pool, Ferry Pool Road, 11A2. *Outdoor facilities are detailed in the section on Parks and Gardens; River Use.*

Notes: On maps 2-15 wide roads are intended only to indicate practical through-routes and internal 'by-passes' for congested areas. Such representation does not imply official approval for these routes.
The representation of a road, or footpath does not imply the existence of a public right of way.
The inclusion of any site or facility in this book does not imply any kind of recommendation or endorsement by the publishers.

MAP LEGEND

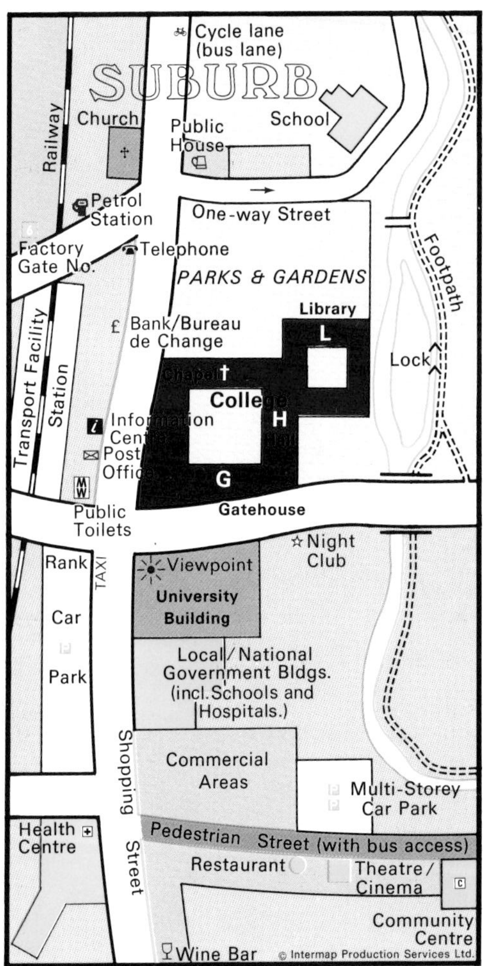
SUBURB
Cycle lane
(bus lane)
Church
Public
House
School
Railway
Petrol
Station
One-way Street
Factory
Gate No.
Telephone
PARKS & GARDENS
Footpath
Library
L
Bank/Bureau
de Change
Lock
Transport Facility
Station
College
H
Information
Centre
Post
Office
G
Public
Toilets
Gatehouse
Night
Club
Rank
TAXI
Viewpoint
University
Building
Car
Park
Local/National
Government Bldgs.
(incl.Schools and
Hospitals.)
Shopping
Street
Commercial
Areas
Multi-Storey
Car Park
Health
Centre
Pedestrian Street (with bus access)
Restaurant
Theatre/
Cinema
Community
Centre
Wine Bar
© Intermap Production Services Ltd.

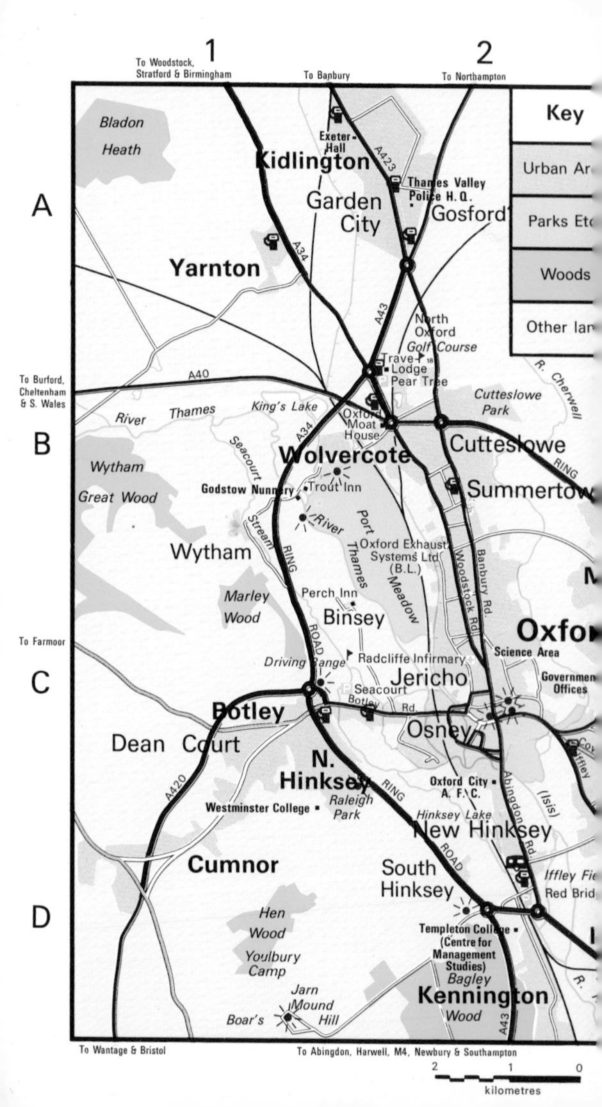
1
2
To Woodstock, Stratford & Birmingham
To Banbury
To Northampton
Key
Urban Ar
Parks Etc
Woods
Other lar
A
B
C
D
To Burford, Cheltenham & S. Wales
To Farmoor
Bladon Heath
Exeter Hall
Kidlington
A423
Thames Valley Police H.Q.
Garden City
Gosford
A34
Yarnton
A43
North Oxford Golf Course
Travel Lodge
Pear Tree
A40
R. Cherwell
Cutteslowe Park
River Thames
King's Lake
Oxford Moat House
Seacourt
A34
Wolvercote
Cutteslowe
RING
Wytham
Great Wood
Godstow Nunnery
Trout Inn
Summertow
Stream
River
Port
Wytham
RING
Oxford Exhaust Systems Ltd (B.L.)
Thames
Meadow
Woodstock Rd.
Banbury Rd.
Marley Wood
Perch Inn
Binsey
Oxfo
ROAD
Science Area
Driving Range
Radcliffe Infirmary
Jericho
Government Offices
Seacourt
Botley
Botley Rd.
Osney
Dean Court
N. Hinksey
RING
A420
Oxford City A. F. C.
Abingdon Rd.
(Isis)
Westminster College
Raleigh Park
Hinksey Lake
New Hinksey
ROAD
Cumnor
South Hinksey
Iffley Fie
Red Brid
Hen Wood
Templeton College (Centre for Management Studies)
Youlbury Camp
Bagley
Jarn Mound
Kennington
Boar's Hill
Wood
A43
To Wantage & Bristol
To Abingdon, Harwell, M4, Newbury & Southampton
2
1
0
kilometres

The City of OXFORD

Dual carriageway
B roads
Railways
A roads
Other roads
Canals

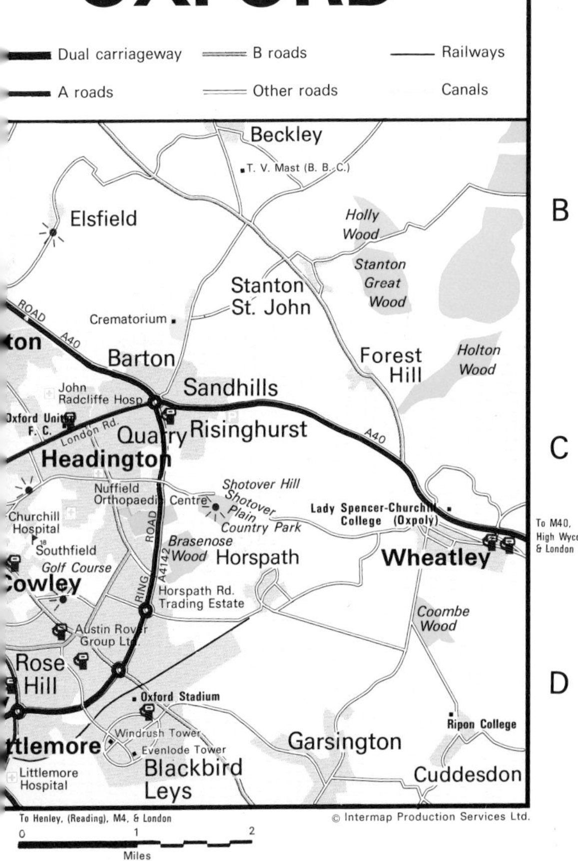

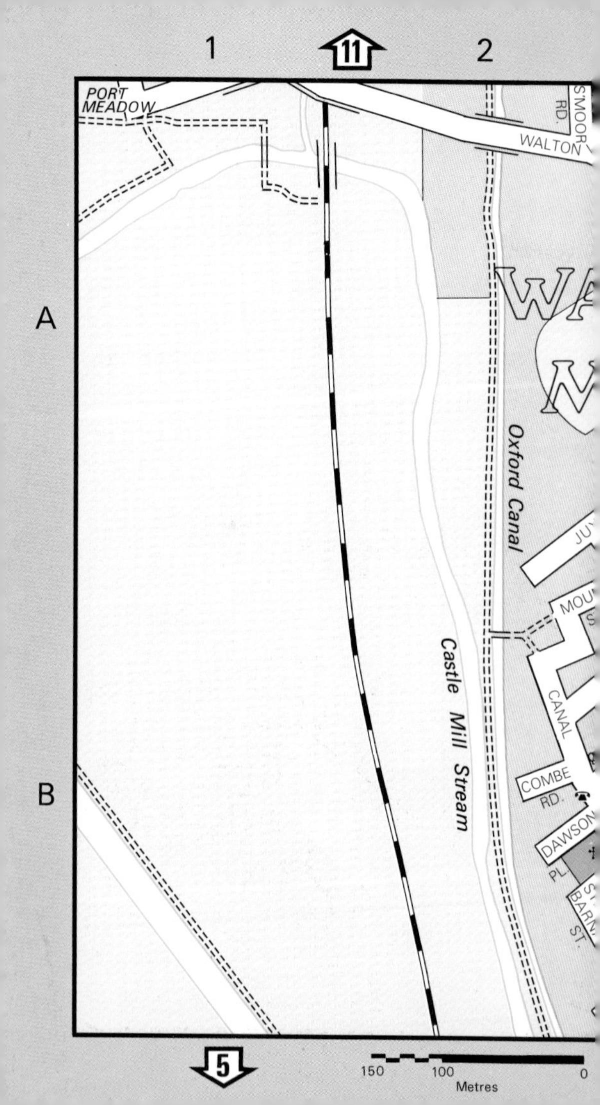
1
11
2
PORT
MEADOW
WALTON
A
Oxford Canal
CANAL
Castle Mill Stream
B
COMBE
RD.
DAWSON
PL.
ST.
5
150
100
0
Metres

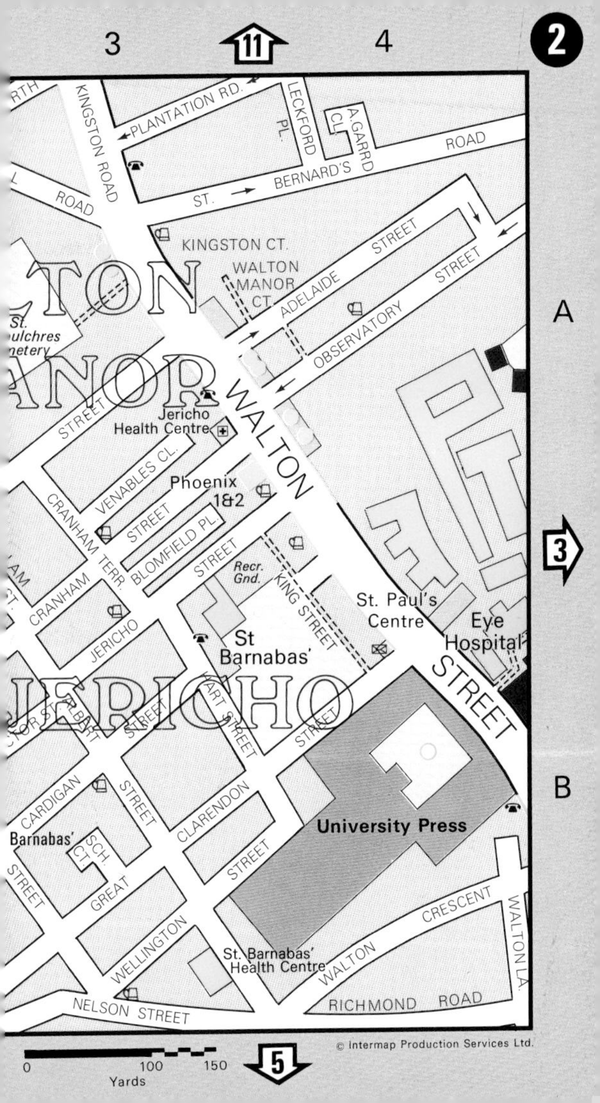
3
11
4
A
B
3
5
PLANTATION RD.
KINGSTON ROAD
LECKFORD PL.
A. GARRD CL.
ROAD
ST. BERNARD'S
ROAD
KINGSTON CT.
WALTON MANOR CT.
ADELAIDE STREET
OBSERVATORY STREET
WALTON MANOR
St. Sepulchres Cemetery
STREET
Jericho Health Centre
WALTON STREET
VENABLES CL.
Phoenix 1&2
CRANHAM TERR.
STREET
BLOMFIELD PL.
STREET
Recr. Gnd.
KING STREET
CRANHAM
JERICHO
St. Paul's Centre
Eye Hospital
St Barnabas'
JERICHO
HART STREET
STREET
STREET
CARDIGAN
STREET
CLARENDON
University Press
Barnabas'
SCH. CT.
STREET
GREAT
STREET
WELLINGTON
St. Barnabas' Health Centre
WALTON CRESCENT
WALTON LA.
NELSON STREET
RICHMOND ROAD
0
100
150
Yards

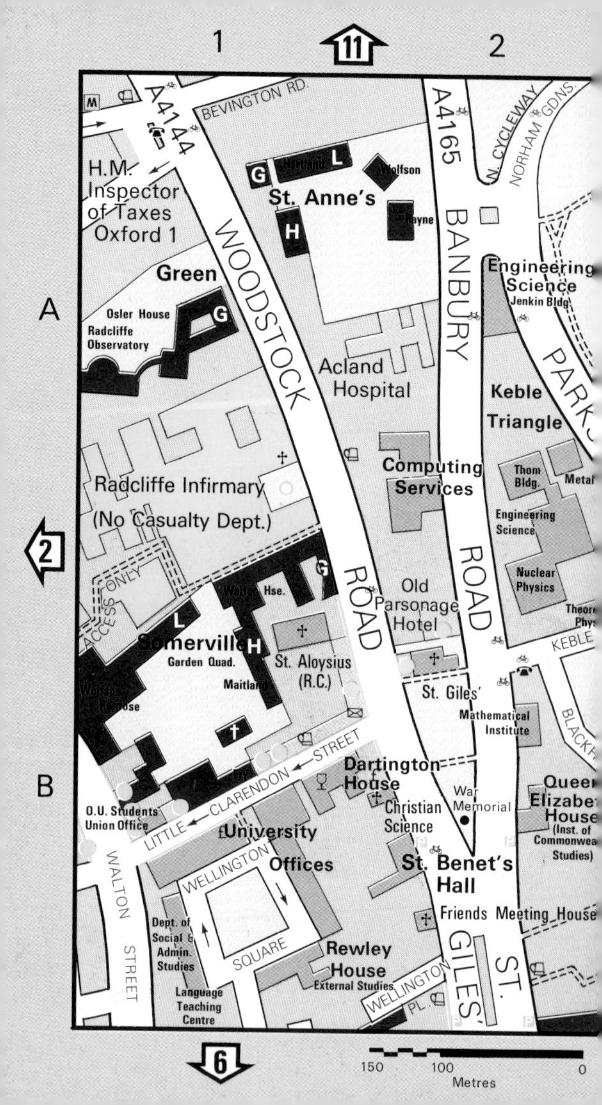

1
11
2
A
B
2
6
BEVINGTON RD.
A4144
A4165
N. CYCLEWAY
NORHAM GDNS.
H.M. Inspector of Taxes Oxford 1
Hartland
Wolfson
St. Anne's
Kayne
Green
Osler House
Radcliffe Observatory
WOODSTOCK ROAD
BANBURY ROAD
Engineering Science
Jenkin Bldg.
PARKS
Acland Hospital
Keble Triangle
Computing Services
Thom Bldg.
Engineering Science
Nuclear Physics
Radcliffe Infirmary
(No Casualty Dept.)
ACCESS ONLY
Walton Hse.
Old Parsonage Hotel
Somerville
Garden Quad.
St. Aloysius (R.C.)
Maitland
St. Giles'
KEBLE
Mathematical Institute
LITTLE CLARENDON STREET
Dartington House
Christian Science
War Memorial
Queen Elizabeth House
(Inst. of Commonwealth Studies)
O.U. Students Union Office
University Offices
St. Benet's Hall
WALTON STREET
WELLINGTON SQUARE
Friends Meeting House
Dept. of Social & Admin. Studies
Rewley House
External Studies
ST. GILES'
Language Teaching Centre
WELLINGTON PL.
150
100
0
Metres

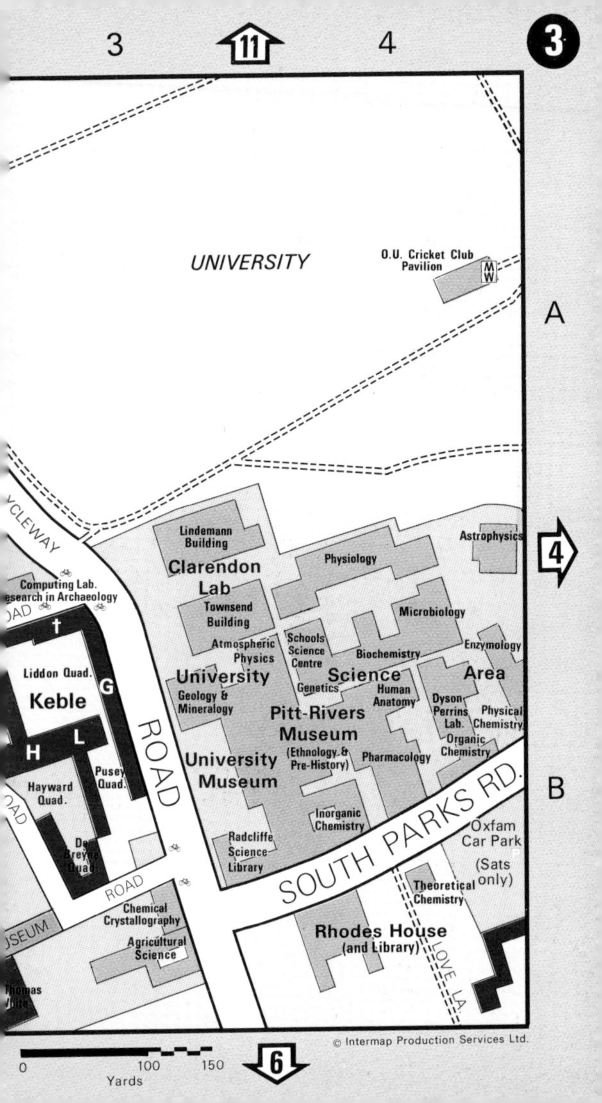
3
11
4
3
UNIVERSITY
O.U. Cricket Club Pavilion
M W
A
4
Lindemann Building
Clarendon Lab
Physiology
Astrophysics
Computing Lab.
Research in Archaeology
Townsend Building
Microbiology
Atmospheric Physics
Schools Science Centre
Biochemistry
Enzymology
Liddon Quad.
G
Keble
University
Science
Area
Genetics
Human Anatomy
Geology & Mineralogy
Dyson Perrins Lab.
Physical Chemistry
Pitt-Rivers Museum
(Ethnology & Pre-History)
Organic Chemistry
L
H
ROAD
University Museum
Pharmacology
Pusey Quad.
B
Hayward Quad.
SOUTH PARKS RD.
Inorganic Chemistry
Oxfam Car Park
(Sats only)
Radcliffe Science Library
ROAD
Theoretical Chemistry
Chemical Crystallography
Rhodes House
(and Library)
Agricultural Science
LOVE LA.
© Intermap Production Services Ltd.
0
100
150
Yards
6

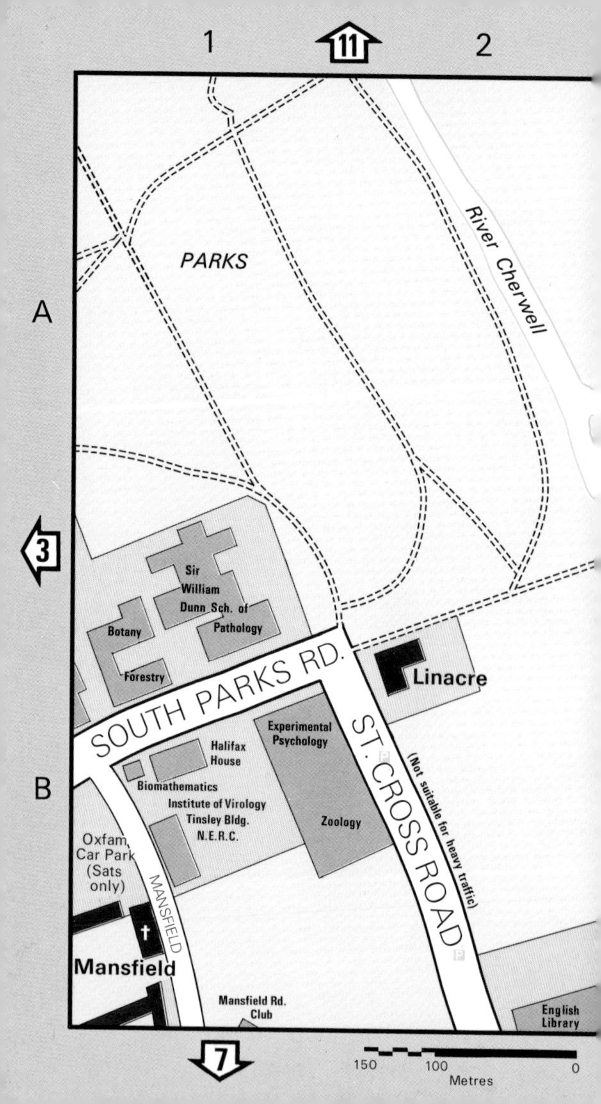
1
11
2
PARKS
River Cherwell
A
3
Sir William Dunn Sch. of Pathology
Botany
Forestry
SOUTH PARKS RD.
Linacre
Experimental Psychology
ST. CROSS ROAD
(Not suitable for heavy traffic)
Halifax House
Biomathematics
Institute of Virology
Tinsley Bldg.
N.E.R.C.
Zoology
B
Oxfam Car Park (Sats only)
MANSFIELD
Mansfield
Mansfield Rd. Club
English Library
7
150
100
0
Metres

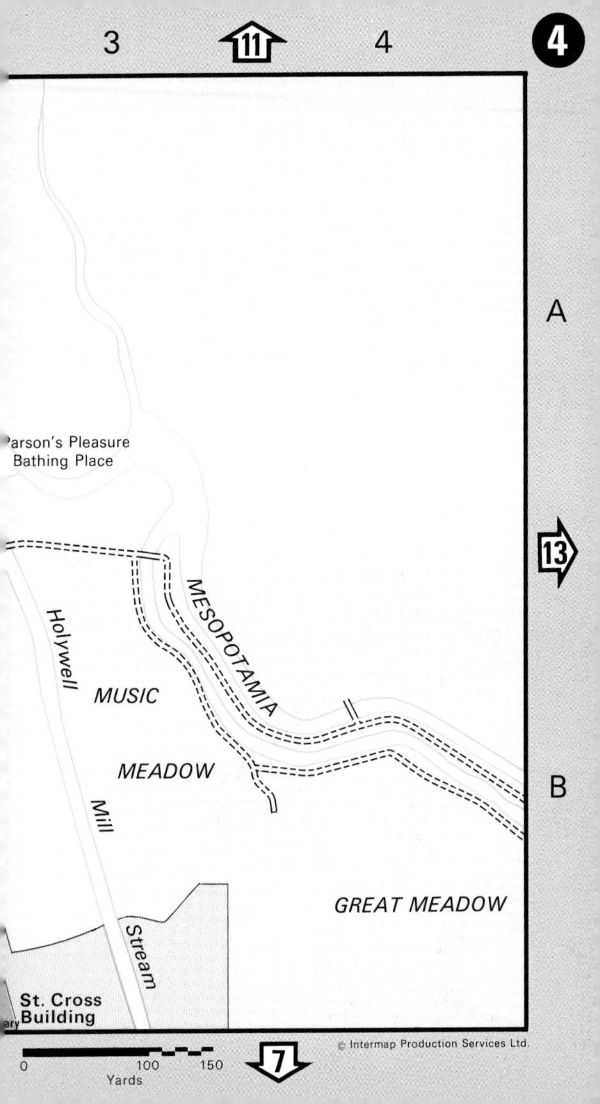
A
B
13
'arson's Pleasure
Bathing Place
MESOPOTAMIA
Holywell
MUSIC
MEADOW
Mill
Stream
GREAT MEADOW
St. Cross
Building
ary
0
100
150
Yards
7

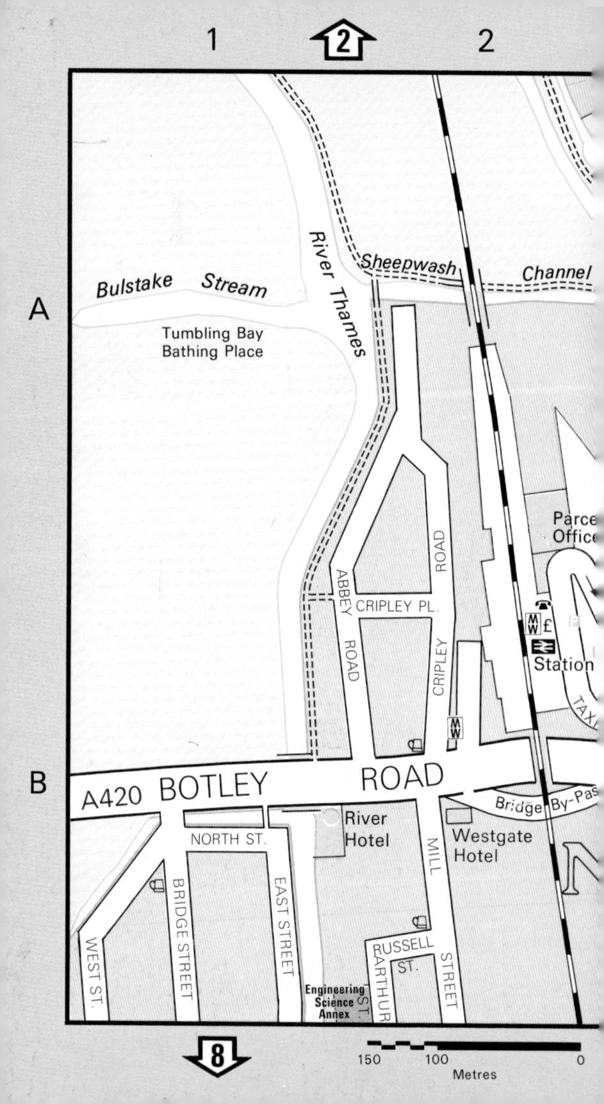
1
2
2
A
B
Bulstake Stream
River Thames
Sheepwash
Channel
Tumbling Bay
Bathing Place
Parce
Office
ABBEY
ROAD
CRIPLEY PL.
ROAD
CRIPLEY
Station
A420 BOTLEY
ROAD
Bridge By-Pas
River
Hotel
Westgate
Hotel
NORTH ST.
MILL
STREET
BRIDGE STREET
EAST STREET
WEST ST.
RUSSELL
ST.
ARTHUR
ST
Engineering
Science
Annex
8
150
100
0
Metres

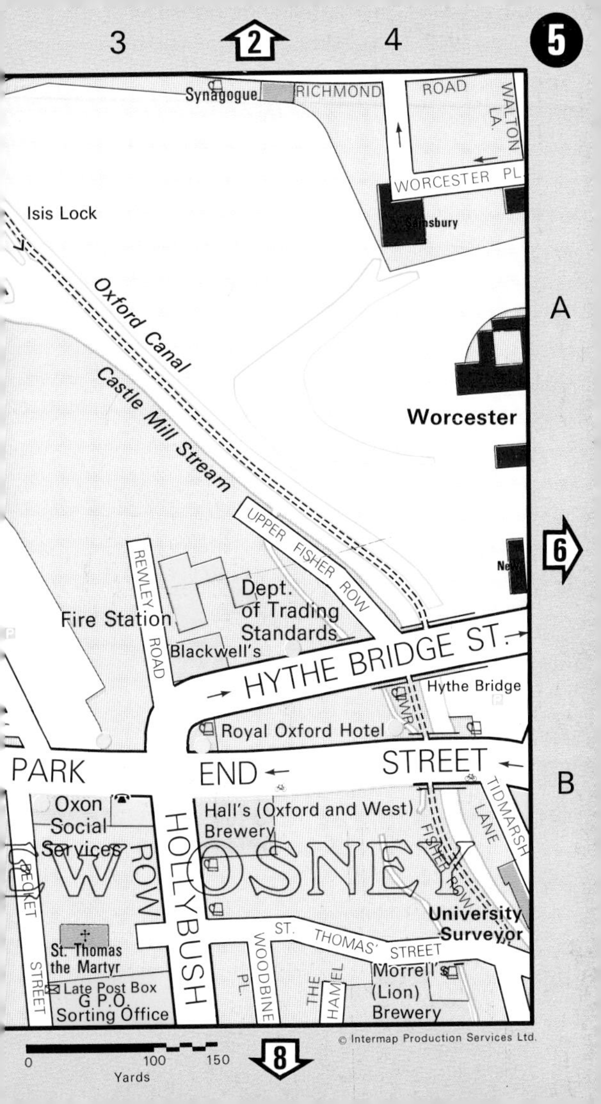
3
2
4
5
Synagogue
RICHMOND
ROAD
WALTON LA.
WORCESTER PL.
Isis Lock
Oxford Canal
Castle Mill Stream
A
Worcester
UPPER FISHER ROW
6
REWLEY ROAD
Fire Station
Dept. of Trading Standards
Blackwell's
HYTHE BRIDGE ST.
Hythe Bridge
Royal Oxford Hotel
PARK
END
STREET
B
Oxon Social Services
Hall's (Oxford and West) Brewery
TIDMARSH LANE
FISHER ROW
OSNEY
BECKET STREET
ROW
HOLLYBUSH
University Surveyor
ST. THOMAS' STREET
St. Thomas the Martyr
Late Post Box
G.P.O. Sorting Office
WOODBINE PL.
THE HAMEL
Morrell's (Lion) Brewery
0
100
150
Yards
8

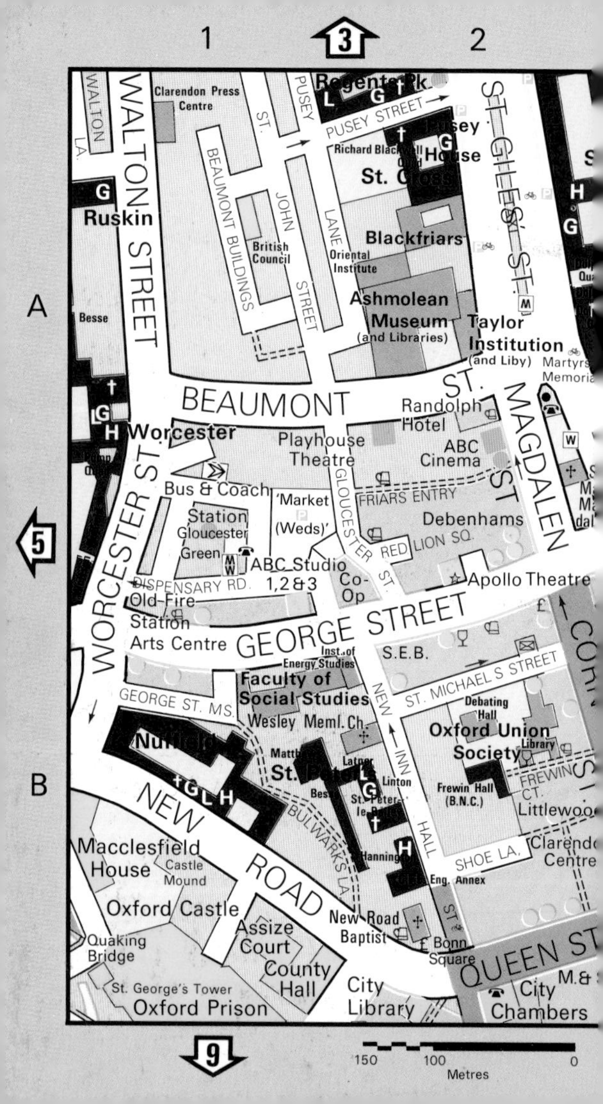
1
3
2
A
B
5
9
WALTON LA.
WALTON STREET
Clarendon Press Centre
PUSEY ST.
Regent's Pk.
PUSEY STREET
Pusey House
Richard Blackwell
St. Cross
ST. GILES' ST.
Ruskin
BEAUMONT BUILDINGS
JOHN STREET
LANE
British Council
Oriental Institute
Blackfriars
Ashmolean Museum
(and Libraries)
Taylor Institution
(and Liby)
Martyrs' Memorial
Besse
BEAUMONT
ST. MAGDALEN ST
Randolph Hotel
Worcester
Playhouse Theatre
ABC Cinema
Bus & Coach
Station
Gloucester Green
'Market (Weds)'
GLOUCESTER ST.
FRIARS ENTRY
Debenhams
RED LION SQ.
ABC Studio 1,2&3
Co-Op
Apollo Theatre
WORCESTER ST.
DISPENSARY RD.
Old Fire Station Arts Centre
GEORGE STREET
S.E.B.
Inst. of Energy Studies
Faculty of Social Studies
Wesley Meml. Ch.
NEW INN HALL ST.
ST. MICHAEL S STREET
Debating Hall
Oxford Union Society
Library
GEORGE ST. MS.
Nuffield
St. Peter's
Linton
St. Peter-le-Bailey
Frewin Hall (B.N.C.)
FREWIN CT.
Littlewoods
NEW ROAD
BULWARKS LA.
Hannington
Eng. Annex
SHOE LA.
Clarendon Centre
Macclesfield House
Castle Mound
Oxford Castle
Assize Court
New Road Baptist
Bonn Square
QUEEN ST
Quaking Bridge
County Hall
St. George's Tower
Oxford Prison
City Library
City Chambers
M.&
150
100
0
Metres

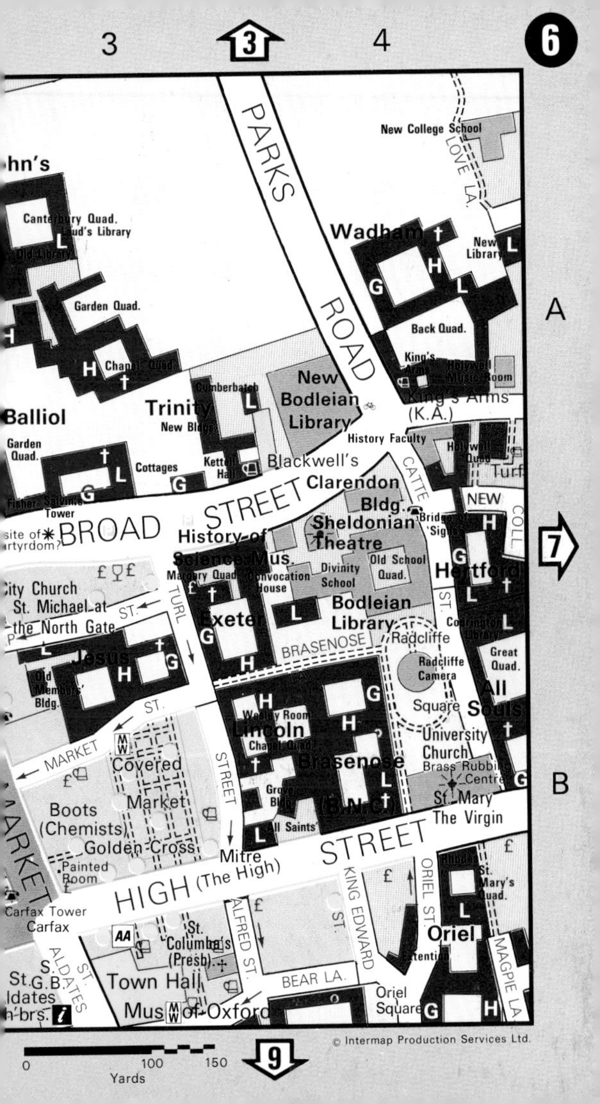
3
3
4
6
PARKS ROAD
New College School
LOVE LA.
Canterbury Quad.
Laud's Library
Wadham
New Library
Garden Quad.
Back Quad.
A
Chapel Quad.
New Bodleian Library
King's Arms
Holywell Music Room
Cumberbatch
King's Arms (K.A.)
Trinity
Balliol
New Bldgs.
History Faculty
Garden Quad.
Cottages
Kettell Hall
Blackwell's
Clarendon Bldg.
CATTE ST.
Holywell Quad
Turf
NEW COLL.
Fisher Salvin Tower
BROAD STREET
Sheldonian Theatre
Bridge of Sighs
site of Martyrdom?
History of Science Mus.
Old School Quad.
Hertford
7
Margary Quad
Convocation House
Divinity School
City Church St. Michael at the North Gate
TURL ST.
Bodleian Library
Codrington Library
Exeter
Radcliffe
Great Quad.
Jesus
BRASENOSE LA.
Radcliffe Camera
Old Members' Bldg.
Wesley Room
All Souls
MARKET ST.
Square
Lincoln
University Church
Chapel Quad.
Brass Rubbing Centre
Covered Market
Brasenose
B
Grove Bldg.
(B.N.C.)
St. Mary The Virgin
Boots (Chemists)
All Saints'
MARKET
Golden Cross
Mitre
STREET
Painted Room
HIGH (The High) STREET
KING EDWARD ST.
ORIEL ST.
Rhodes
St. Mary's Quad.
Carfax Tower
Carfax
ALFRED ST.
Oriel
St. Columba's (Presb).
MAGPIE LA.
ST. ALDATES
Town Hall
BEAR LA.
Oriel Square
St. Aldates
Mus of Oxford
© Intermap Production Services Ltd.
0 100 150 Yards
9

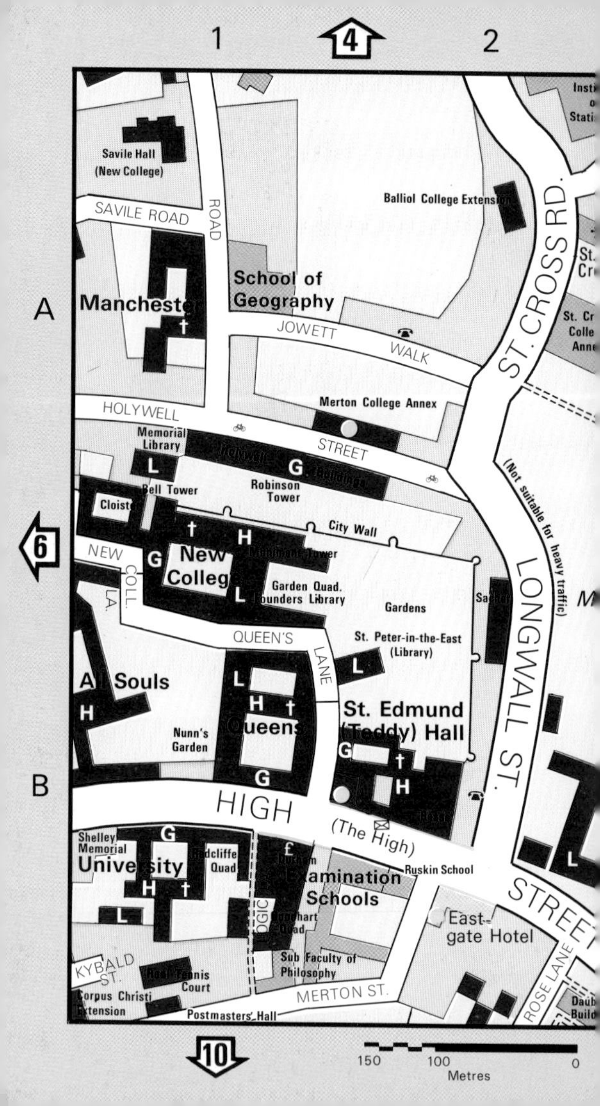
1
4
2
A
B
6
10
Savile Hall
(New College)
SAVILE ROAD
ROAD
Balliol College Extension
School of
Geography
Manchester
JOWETT
WALK
ST. CROSS RD.
Merton College Annex
HOLYWELL
STREET
Memorial
Library
Robinson
Tower
Bell Tower
Cloisters
City Wall
NEW
New
College
COLL.
LA.
Garden Quad.
Founders Library
Gardens
QUEEN'S
LANE
St. Peter-in-the-East
(Library)
LONGWALL ST.
(Not suitable for heavy traffic)
All Souls
Queens
St. Edmund
(Teddy) Hall
Nunn's
Garden
HIGH
(The High)
Shelley
Memorial
University
Radcliffe
Quad
Examination
Schools
Ruskin School
STREET
East-
gate Hotel
LOGIC
Sub Faculty of
Philosophy
KYBALD
ST.
Real Tennis
Court
Corpus Christi
Extension
MERTON ST.
Postmasters' Hall
ROSE LANE
150
100
0
Metres

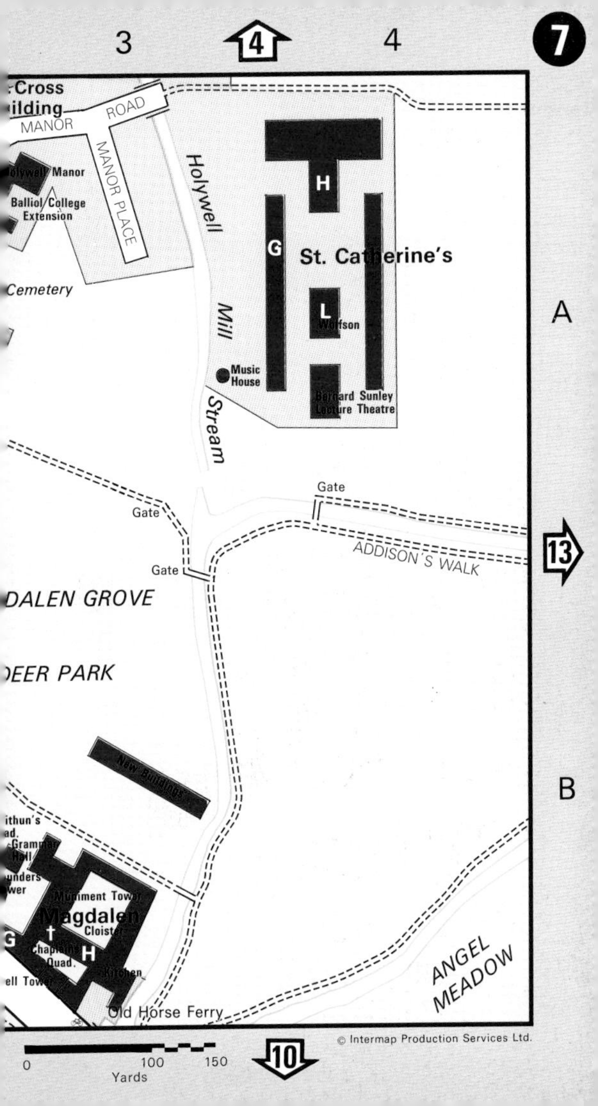
3
4
4
7
Cross
ilding
MANOR ROAD
MANOR PLACE
Holywell Manor
Balliol College
Extension
Holywell
Mill
Stream
St. Catherine's
H
G
L
Wolfson
Music
House
Bernard Sunley
Lecture Theatre
Cemetery
A
Gate
Gate
Gate
ADDISON'S WALK
13
DALEN GROVE
DEER PARK
New Buildings
B
Grammar
Hall
Muniment Tower
Magdalen
Cloister
Chaplains
Quad.
Kitchen
H
G
Old Horse Ferry
ANGEL
MEADOW
0
100
150
Yards
10

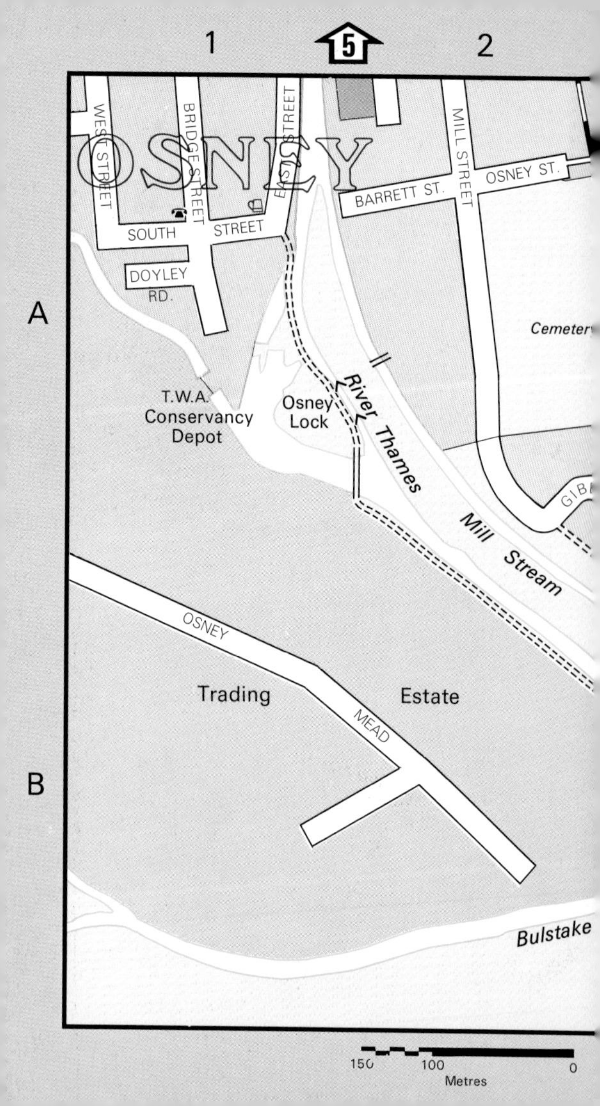
1
5
2
OSNEY
WEST STREET
BRIDGE STREET
EAST STREET
MILL STREET
OSNEY ST.
BARRETT ST.
SOUTH
STREET
DOYLEY
RD.
A
Cemetery
T.W.A.
Conservancy
Depot
Osney
Lock
River Thames
Mill
Stream
OSNEY
Trading
Estate
MEAD
B
Bulstake
150
100
0
Metres

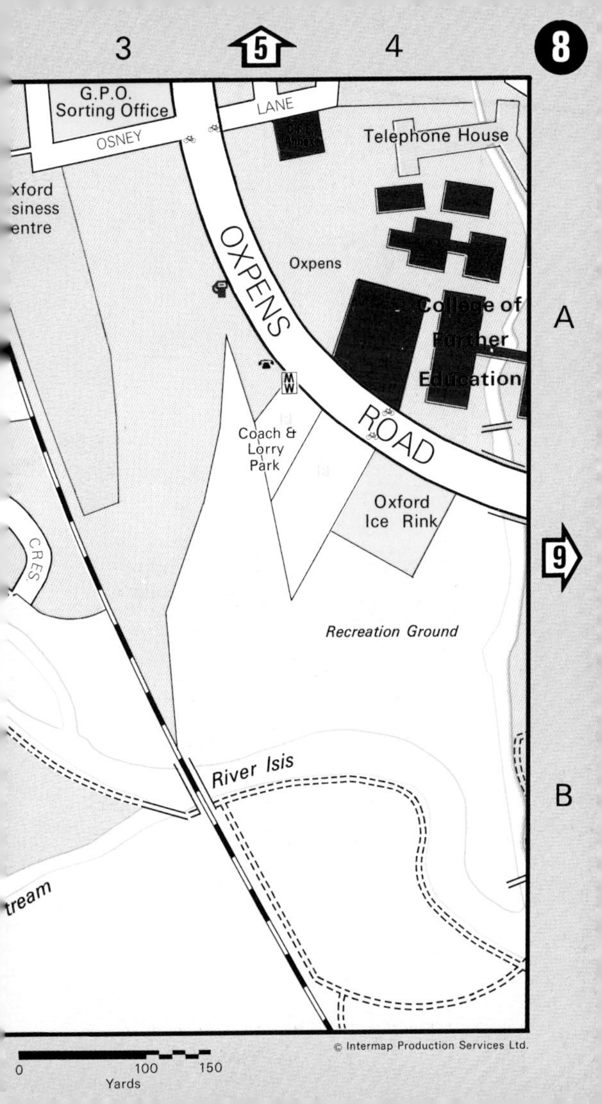
G.P.O.
Sorting Office
OSNEY
LANE
Telephone House
xford
siness
entre
OXPENS
Oxpens
College of
Further
Education
ROAD
Coach &
Lorry
Park
Oxford
Ice Rink
C.RES.
Recreation Ground
River Isis
tream
A
B
9
0
100
150
Yards

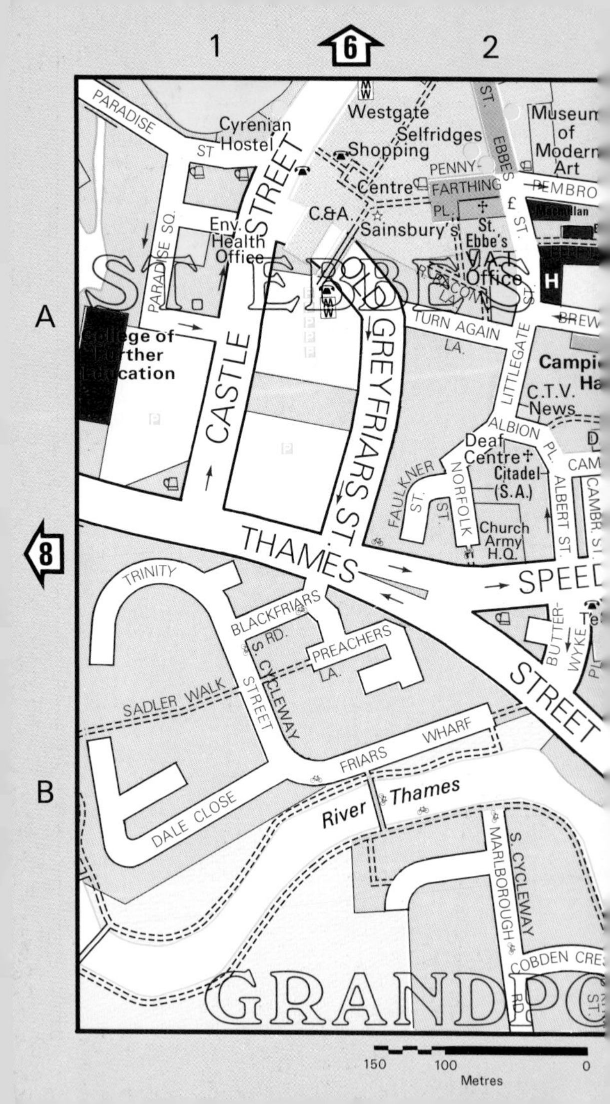

1
6
2
A
B
8
PARADISE ST
Cyrenian Hostel
Westgate
Selfridges
Shopping
Centre
Museum of Modern Art
PENNY-FARTHING PL.
ST. EBBES ST.
PEMBRO
C.&A.
Sainsbury's
St. Ebbe's
Env. Health Office
V.A.T. Office
PARADISE SQ.
ST. EBBES
TURN AGAIN LA.
BREW
College of Further Education
CASTLE STREET
GREYFRIARS ST.
LITTLEGATE ST.
C.T.V. News
ALBION PL.
Deaf Centre
Citadel (S.A.)
FAULKNER ST.
NORFOLK ST.
ALBERT ST.
Church Army H.Q.
THAMES
SPEED
STREET
TRINITY
BLACKFRIARS RD.
PREACHERS LA.
BUTTER-WYKE
S. CYCLEWAY
SADLER WALK
STREET
WHARF
FRIARS
River Thames
DALE CLOSE
MARLBOROUGH RD.
S. CYCLEWAY
COBDEN CRES
GRANDPO
150
100
0
Metres

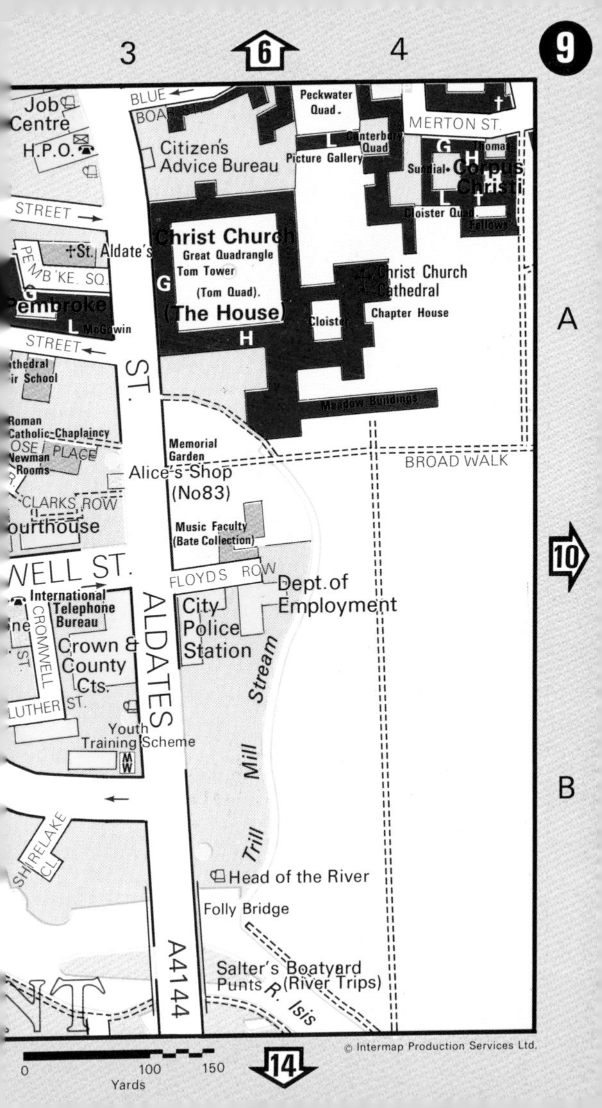

3
6
4
9
Job Centre
H.P.O.
BLUE BOAR ST.
Peckwater Quad.
MERTON ST.
Canterbury Quad
Citizen's Advice Bureau
Picture Gallery
Thomas
Sundial
Corpus Christi
Cloister Quad.
Fellows
STREET
Christ Church
St. Aldate's
Great Quadrangle
Tom Tower
PEMB'KE. SQ.
(Tom Quad).
(The House)
Christ Church Cathedral
Chapter House
Cloister
Pembroke
McGowin
STREET
A
Cathedral Choir School
ST.
Meadow Buildings
Roman Catholic Chaplaincy
ROSE PLACE
Newman Rooms
Memorial Garden
BROAD WALK
Alice's Shop (No83)
CLARKS ROW
Courthouse
Music Faculty (Bate Collection)
10
WELL ST.
FLOYDS ROW
Dept. of Employment
International Telephone Bureau
City Police Station
CROMWELL ST.
Crown & County Cts.
ALDATES
Stream
LUTHER ST.
Youth Training Scheme
Mill
B
SHIRELAKE CL.
Trill
Head of the River
Folly Bridge
A4144
Salter's Boatyard (River Trips)
Punts
R. Isis
© Intermap Production Services Ltd.
0
100
150
Yards
14

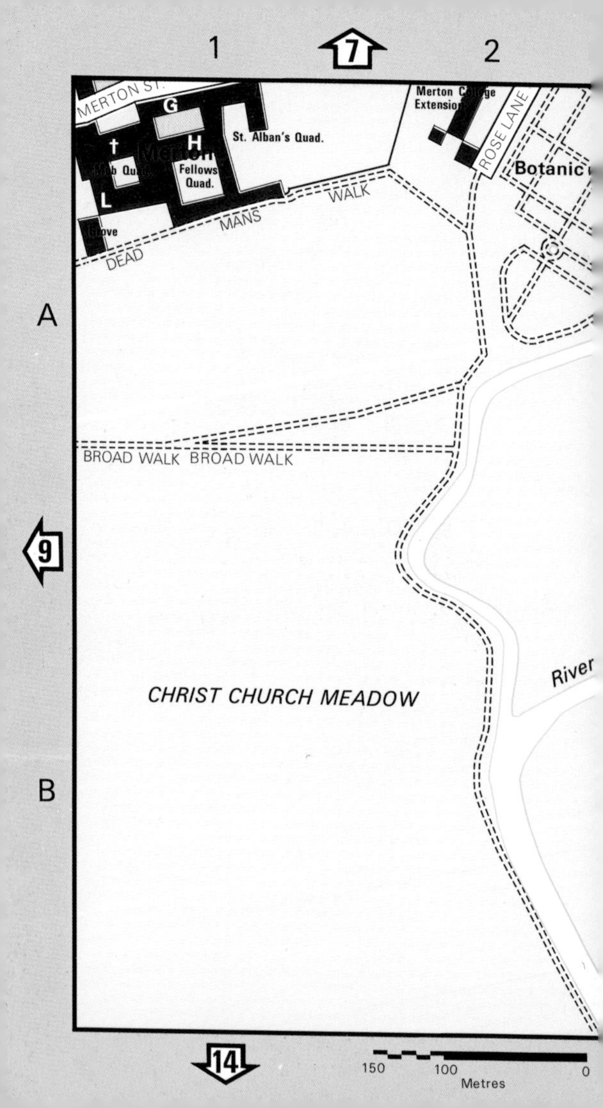
1
7
2
MERTON ST.
G
H
St. Alban's Quad.
Fellows
Quad.
L
DEAD
MANS
WALK
Merton College
Extension
ROSE LANE
Botanic
A
BROAD WALK
BROAD WALK
9
River
CHRIST CHURCH MEADOW
B
14
150
100
Metres
0

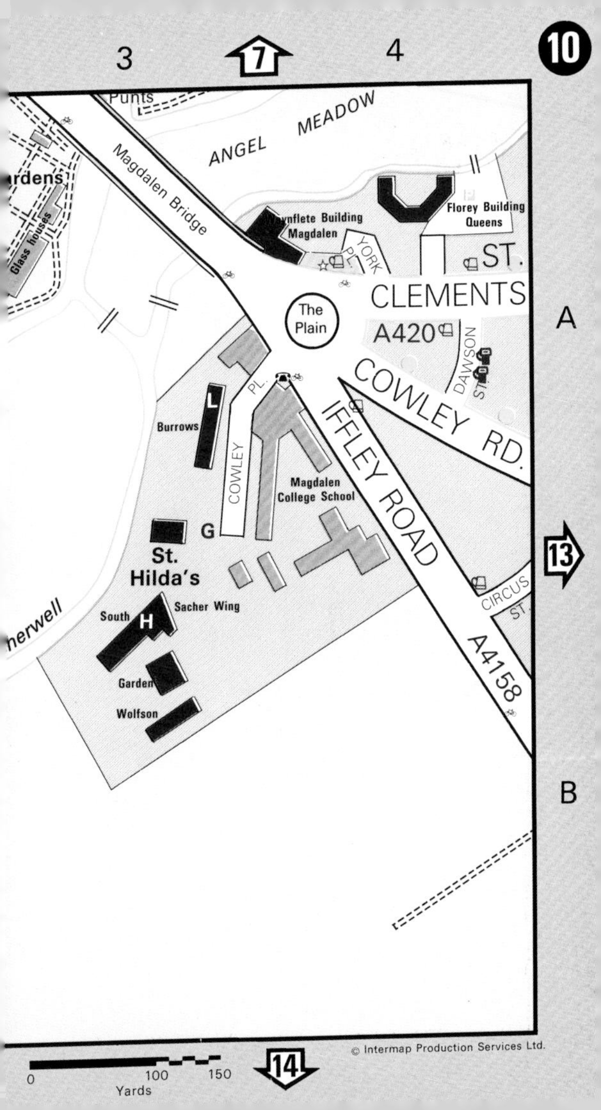
3
7
4
Punts
ANGEL MEADOW
Magdalen Bridge
ardens
Glass houses
Waynflete Building
Magdalen
Florey Building
Queens
YORK PL.
ST. CLEMENTS
A420
The Plain
A
DAWSON ST.
COWLEY RD.
COWLEY PL.
L
Burrows
IFFLEY ROAD
Magdalen College School
G
St. Hilda's
Sacher Wing
South
H
Garden
Wolfson
herwell
CIRCUS ST.
A4158
13
B
14
0
100
150
Yards

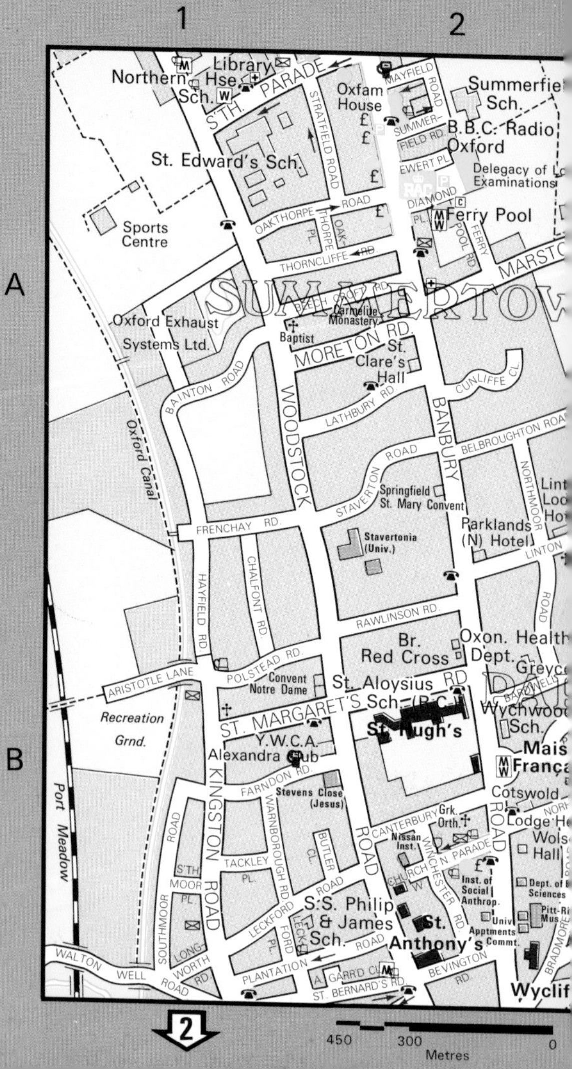

1 2

A

B

Library
Northern Hse Sch.
S'TH. PARADE
Oxfam House
MAYFIELD ROAD
Summerfie[ld] Sch.
STRATFIELD ROAD
SUMMERFIELD RD.
B.B.C. Radio Oxford
EWERT PL.
Delegacy of Lo[cal] Examinations
St. Edward's Sch.
RAC
DIAMOND PL.
OAKTHORPE ROAD
OAKTHORPE PL.
Ferry Pool
FERRY POOL RD.
Sports Centre
THORNCLIFFE RD.
MARSTO[N]
SUMMERTOWN
BEECH CROFT RD.
Carmelite Monastery
Oxford Exhaust Systems Ltd.
Baptist
MORETON RD.
St. Clare's Hall
BAINTON ROAD
CUNLIFFE CL.
Oxford Canal
WOODSTOCK
LATHBURY RD.
BANBURY
BELBROUGHTON ROA[D]
ROAD
STAVERTON
Springfield St. Mary Convent
NORTHMOOR ROAD
Lint[on] Lod[ge] Ho[tel]
FRENCHAY RD.
Parklands (N) Hotel
Stavertonia (Univ.)
LINTON
CHALFONT RD.
HAYFIELD RD.
RAWLINSON RD.
Br. Red Cross
Oxon. Health Dept.
POLSTEAD RD.
Greyc[otes]
ARISTOTLE LANE
Convent Notre Dame
St. Aloysius Sch. (R.C.)
RD
BARDWELL
PA[RK]
Recreation Grnd.
ST. MARGARET'S
Wychwoo[d] Sch.
St. Hugh's
Y.W.C.A.
Alexandra Club
Mais[on] França[ise]
Port Meadow
KINGSTON ROAD
FARNDON RD.
Stevens Close (Jesus)
Cotswold
WARNBOROUGH RD.
ROAD
Lodge H[otel]
CANTERBURY
Grk. Orth.
NOR[HAM]
ROAD
BUTLER CL.
Nissan Inst.
N. PARADE
Wols[ey] Hall
TACKLEY PL.
CHURCH W
WINCHESTER RD.
S'TH MOOR PL.
Inst. of Social Anthrop.
Dept. of [Earth] Sciences
S.S. Philip & James Sch.
LECKFORD ROAD
Pitt-Ri[vers] Mus.
SOUTHMOOR
LECKFORD PL.
St. Anthony's
Univ. Apptments Commt.
LONGWORTH RD.
ROAD
BRADMORE
WALTON WELL ROAD
PLANTATION
GARR'D CL.
BEVINGTON RD.
ST. BERNARD'S RD.
Wyclif[fe]

2

450 300 0
Metres

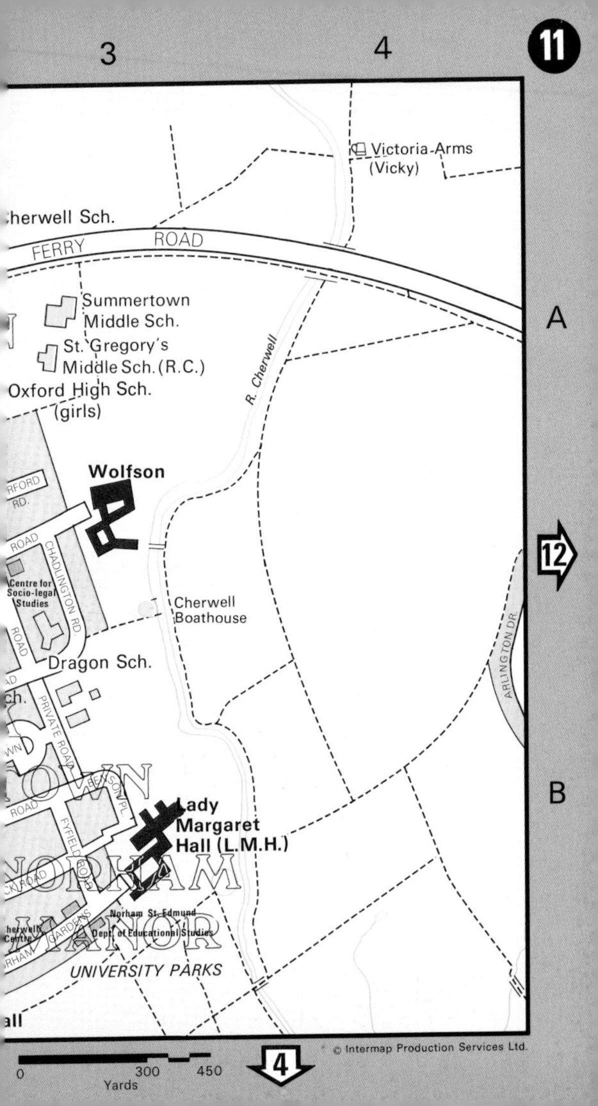
3
4
Victoria Arms
(Vicky)
Cherwell Sch.
FERRY
ROAD
Summertown
Middle Sch.
St. Gregory's
Middle Sch. (R.C.)
Oxford High Sch.
(girls)
R. Cherwell
Wolfson
RD.
ROAD
CHADLINGTON RD.
Centre for
Socio-legal
Studies
ROAD
Cherwell
Boathouse
Dragon Sch.
PRIVATE ROAD
ARLINGTON DR.
A
B
12
BENSON PL.
ROAD
FYFIELD ROAD
Lady
Margaret
Hall (L.M.H.)
NORHAM
MANOR
Norham St. Edmund
Dept. of Educational Studies
GARDENS
UNIVERSITY PARKS
4
0
300
450
Yards

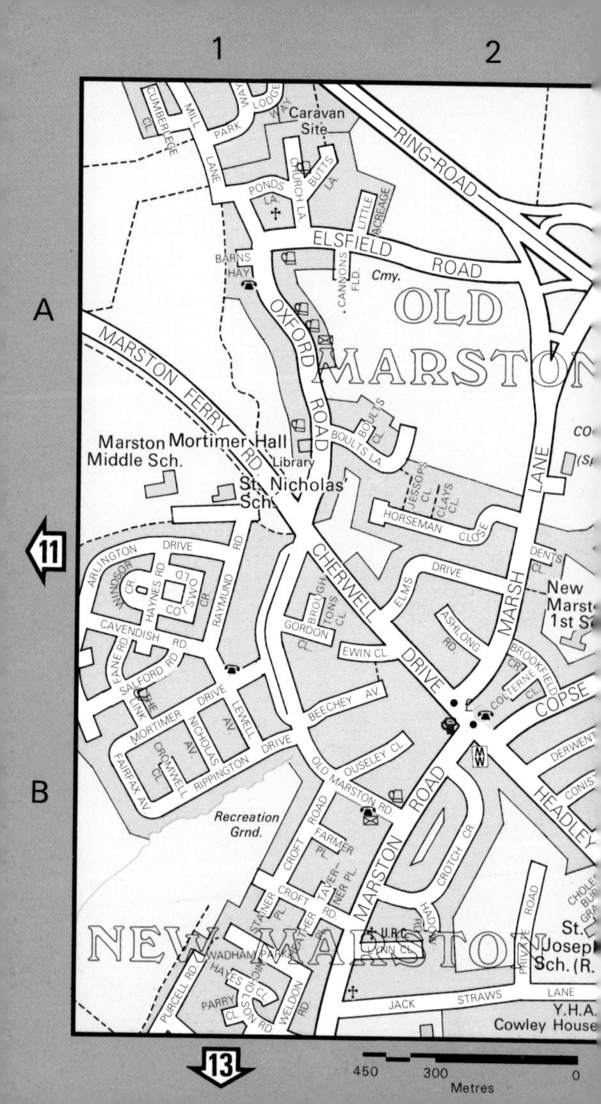
1
2
A
B
11
13
CUMBERLEGE CL.
MILL LANE
PARK
WAY
LODGE
WAY
Caravan Site
RING-ROAD
PONDS LA.
CHURCH LA.
BUTTS LA.
LITTLE ACREAGE
ELSFIELD
ROAD
BARNS HAY
CANNONS FLD.
Cmy.
OLD MARSTON
OXFORD ROAD
MARSTON FERRY RD
BOULTS CL.
BOULTS LA.
Marston Middle Sch.
Mortimer Hall
Library
St. Nicholas' Sch.
JESSOPS CL.
CLAYS CL.
HORSEMAN CLOSE
MARSH LANE
DENTS CL.
New Marst 1st S
ARLINGTON DRIVE
WINDSOR CR.
HAYNES RD.
COTSWOLD CR.
RAYMUND RD.
CHERWELL DRIVE
BROUGHTONS CL.
GORDON CL.
ELMS DRIVE
ASHLONG RD.
CAVENDISH RD.
FANE RD
SALFORD RD.
THE LINK
EWIN CL.
BROOKFIELD CR.
COLTERNE CL.
COPSE
MORTIMER DRIVE
NICHOLAS AV.
LEWELL AV.
BEECHEY AV
DERWENT
FAIRFAX AV
CROMWELL CL.
RIPPINGTON DRIVE
OLD MARSTON RD.
OUSELEY CL.
HEADLEY
CONIST
Recreation Grnd.
CROFT ROAD
FARMER PL.
MARSTON ROAD
CROTCH CR.
CROFT RD.
TAVERNER PL.
STAINER PL.
HADOW RD.
PRIVATE ROAD
St. Joseph Sch. (R.
NEW MARSTON
U.R.C.
LYNN CL.
WADHAM PARK
PURCELL RD.
HAYES CL.
GIBBONS
PARRY CL.
WELDON RD.
JACK STRAWS LANE
Y.H.A. Cowley House
450
300
0
Metres

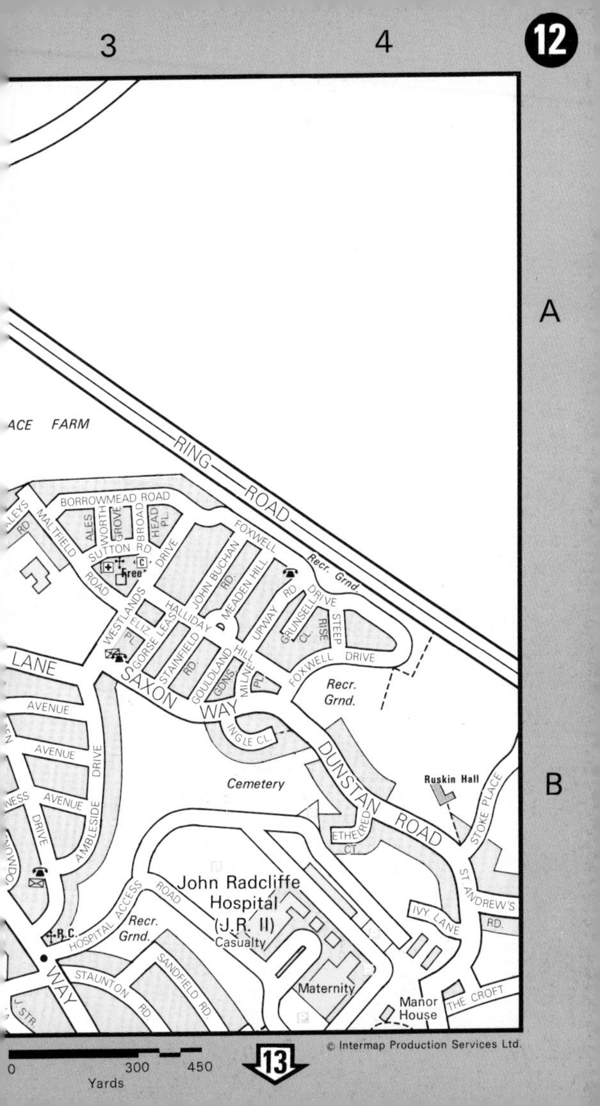
A
B
ACE FARM
RING ROAD
BORROWMEAD ROAD
MALTFIELD RD
ALES
WORTH
GROVE
BROAD
HEAD PL.
SUTTON RD
DRIVE
ROAD
FOXWELL
Recr. Grnd.
JOHN BUCHAN RD.
MEADEN HILL
UPWAY RD.
GRUNSELL CL.
RISE
STEEP
FOXWELL DRIVE
WESTLANDS
HALLIDAY
ELIZ. PL
GORSE LEAS
STAINFIELD RD.
GOULDLAND GDNS
HILL
MILNE PL
FOXWELL DRIVE
Recr. Grnd.
LANE
SAXON WAY
INGLE CL.
AVENUE
AVENUE
AVENUE
DRIVE
AMBLESIDE
DRIVE
Cemetery
DUNSTAN ROAD
Ruskin Hall
STOKE PLACE
ETHELRED CT.
John Radcliffe Hospital (J.R. II)
Casualty
ROAD
HOSPITAL ACCESS
Recr. Grnd.
R.C.
ST. ANDREW'S RD.
IVY LANE
WAY
STAUNTON RD.
SANDFIELD RD.
Maternity
Manor House
THE CROFT
0
300
450
Yards

13

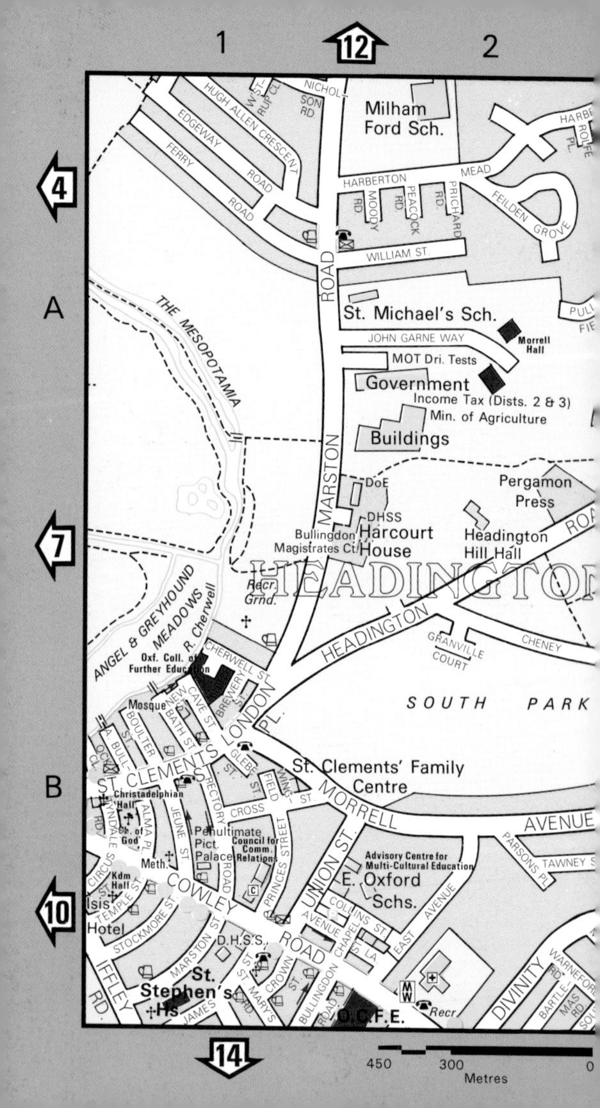

1
12
2
4
7
10
14
A
B
NICHOLSON RD
ST. RUP CL
HUGH ALLEN CRESCENT
EDGEWAY ROAD
FERRY ROAD
Milham Ford Sch.
HARBERTON MEAD
MOODY RD.
PEACOCK RD.
PRICHARD RD.
FEILDEN GROVE
ROLFE PL.
WILLIAM ST.
THE MESOPOTAMIA
St. Michael's Sch.
JOHN GARNE WAY
Morrell Hall
MOT Dri. Tests
Government
Income Tax (Dists. 2 & 3)
Min. of Agriculture
Buildings
MARSTON ROAD
DoE
Pergamon Press
DHSS
Bullingdon Magistrates Ct.
Harcourt House
Headington Hill Hall
HEADINGTON
Recr. Grnd.
ANGEL & GREYHOUND MEADOWS
R. Cherwell
Oxf. Coll. of Further Education
CHERWELL ST.
HEADINGTON ROAD
GRANVILLE COURT
CHENEY
SOUTH PARK
LONDON PL.
Mosque
NEW CAVE ST.
BATH ST.
BREWERY ST.
BOULTER ST.
GLEBE ST.
ST. CLEMENTS ST.
St. Clements' Family Centre
MORRELL AVENUE
Christadelphian Hall
CROSS ST.
RECTORY ROAD
FIELD ST.
WINGFIELD ST.
Ch. of God
ALMA PL.
JEUNE ST.
Penultimate Pict. Palace
Council for Comm. Relations
PRINCES STREET
UNION ST.
Advisory Centre for Multi-Cultural Education
PARSONS PL.
TAWNEY ST.
Meth.
Kdm Hall
CIRCUS ST.
TEMPLE ST.
Isis Hotel
COWLEY ROAD
E. Oxford Schs.
COLLINS ST.
AVENUE
EAST AVENUE
STOCKMORE ST.
MARSTON ST.
D.H.S.S.
CHAPEL ST.
ST. MARY'S RD.
CROWN ST.
BULLINGDON ROAD
IFFLEY RD.
St. Stephen's Hs.
JAMES ST.
O.C.F.E.
Recr.
DIVINITY RD.
WARNEFORD RD.
BARTLEMAS RD.
450
300
0
Metres

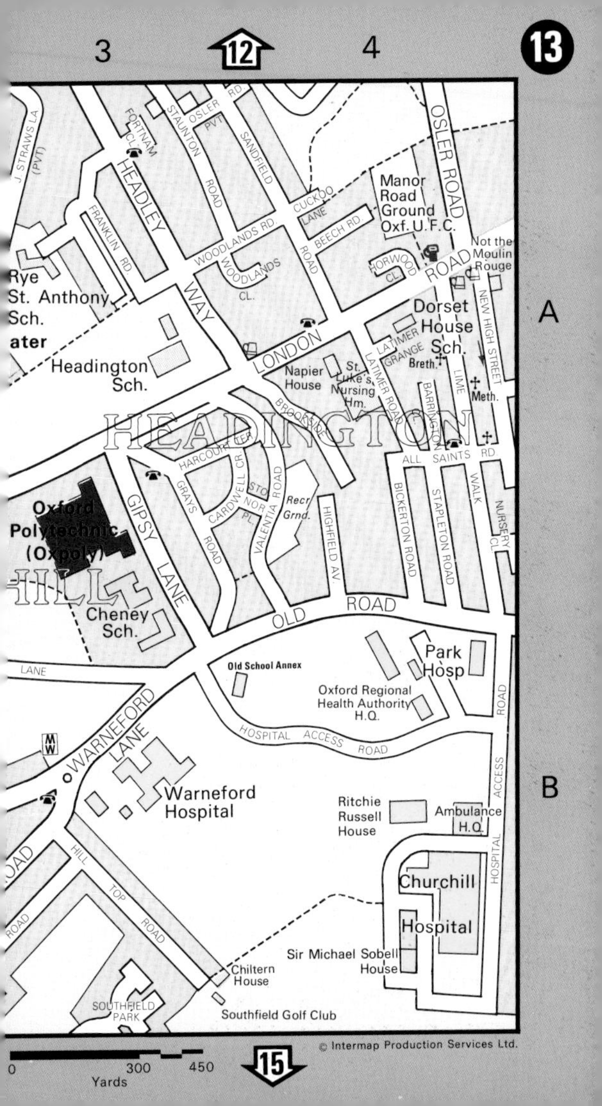
3
12
4
A
B
J. STRAWS LA. (PVT)
FORTNAM CL.
STAUNTON ROAD
OSLER PVT.
OSLER RD.
SANDFIELD ROAD
HEADLEY WAY
FRANKLIN RD.
WOODLANDS RD.
WOODLANDS CL.
CUCKOO LANE
BEECH RD.
Manor Road Ground Oxf. U.F.C.
OSLER ROAD
Not the Moulin Rouge
HORWOOD CL.
ROAD
Rye St. Anthony Sch.
ater
Headington Sch.
LONDON
Dorset House Sch.
LATIMER GRANGE
Breth.
NEW HIGH STREET
Napier House
St. Luke's Nursing Hm.
LATIMER ROAD
LIME WALK
Meth.
BARRINGTON
BROOKSIDE
HEADINGTON
HARCOURT TER.
ALL SAINTS RD.
GRAYS ROAD
CARDWELL CR.
STONOR PL.
VALENTIA ROAD
Recr. Grnd.
HIGHFIELD AV.
BICKERTON ROAD
STAPLETON ROAD
NURSERY CL.
Oxford Polytechnic (Oxpoly)
GIPSY LANE
HILL
Cheney Sch.
OLD ROAD
Park Hosp
Old School Annex
LANE
Oxford Regional Health Authority H.Q.
ROAD
WARNEFORD LANE
HOSPITAL ACCESS ROAD
MW
Warneford Hospital
Ritchie Russell House
Ambulance H.Q.
ACCESS
HOSPITAL
ROAD
HILL TOP ROAD
Churchill Hospital
Sir Michael Sobell House
Chiltern House
SOUTHFIELD PARK
Southfield Golf Club
0
300
450
Yards
15

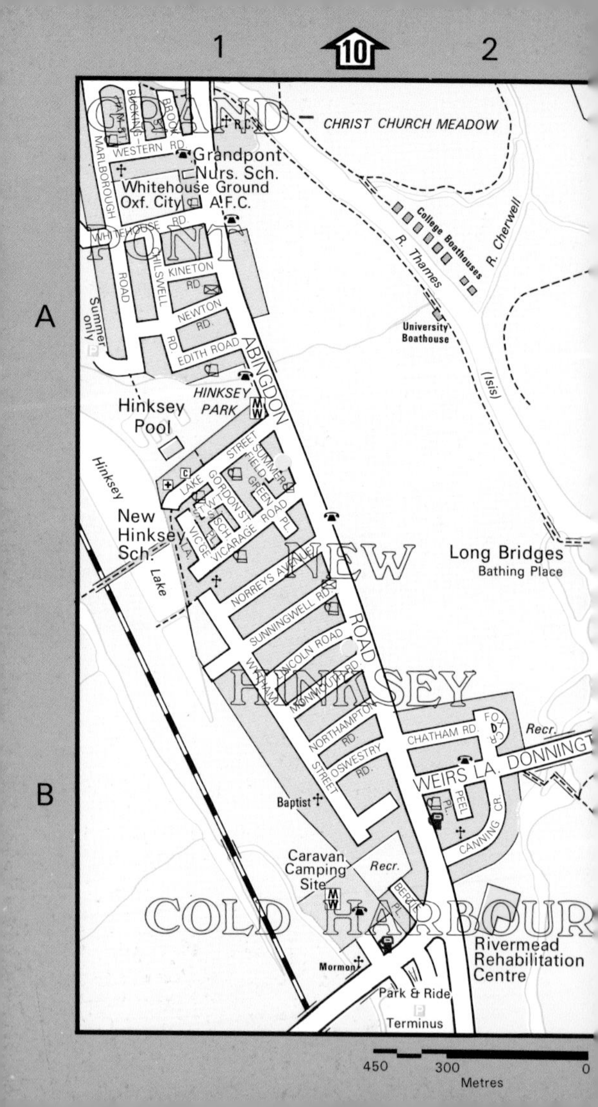

1
10
2
A
B
GRANDPONT
NEW HINKSEY
COLD HARBOUR
CHRIST CHURCH MEADOW
College Boathouses
R. Thames
R. Cherwell
University Boathouse
(Isis)
Grandpont Nurs. Sch.
Whitehouse Ground Oxf. City A.F.C.
R.C.
WESTERN RD.
WHITEHOUSE RD.
MARLBOROUGH ROAD
BUCKINGHAM ST.
BROOK ST.
CHILSWELL RD.
KINETON RD.
NEWTON RD.
EDITH ROAD
ABINGDON ROAD
Summer only
HINKSEY PARK
Hinksey Pool
Hinksey Lake
New Hinksey Sch.
LAKE STREET
GORDON ST.
SUMMERFIELD
GREEN PL.
VICARAGE ROAD
VIC'GE LA.
NORREYS AVENUE
SUNNINGWELL RD.
LINCOLN ROAD
MONMOUTH RD.
NORTHAMPTON RD.
OSWESTRY RD.
WYTHAM STREET
Long Bridges Bathing Place
CHATHAM RD.
FOX CR.
Recr.
DONNINGT
WEIRS LA.
PEEL PL.
CANNING CR.
Baptist
Caravan Camping Site
Recr.
BERKLEY PL.
Mormon
Rivermead Rehabilitation Centre
Park & Ride Terminus
450
300
0
Metres

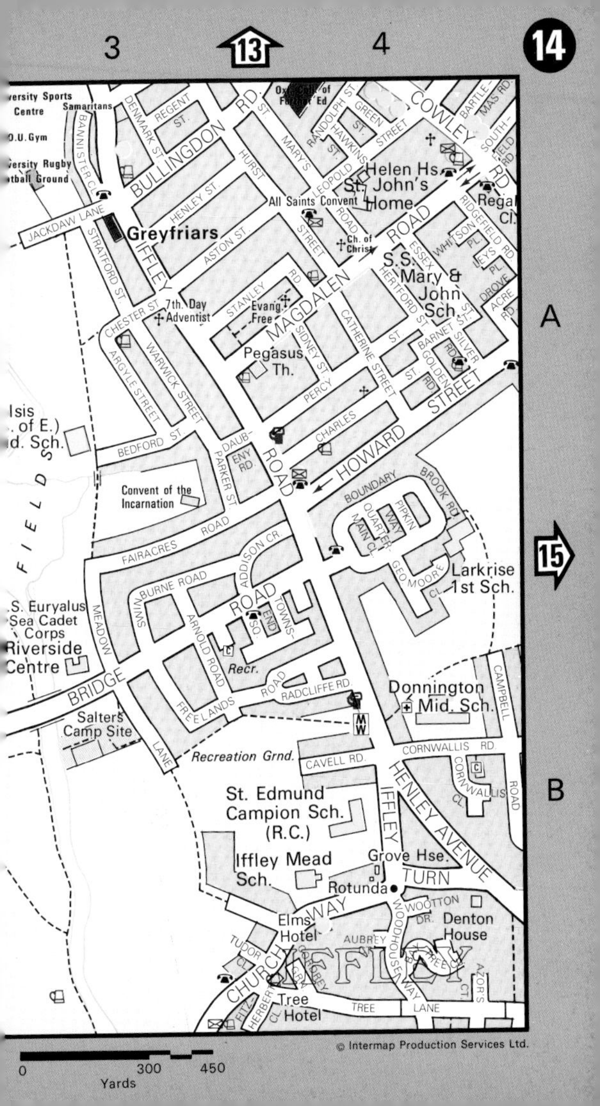

15

A

B

0 300 450 Yards

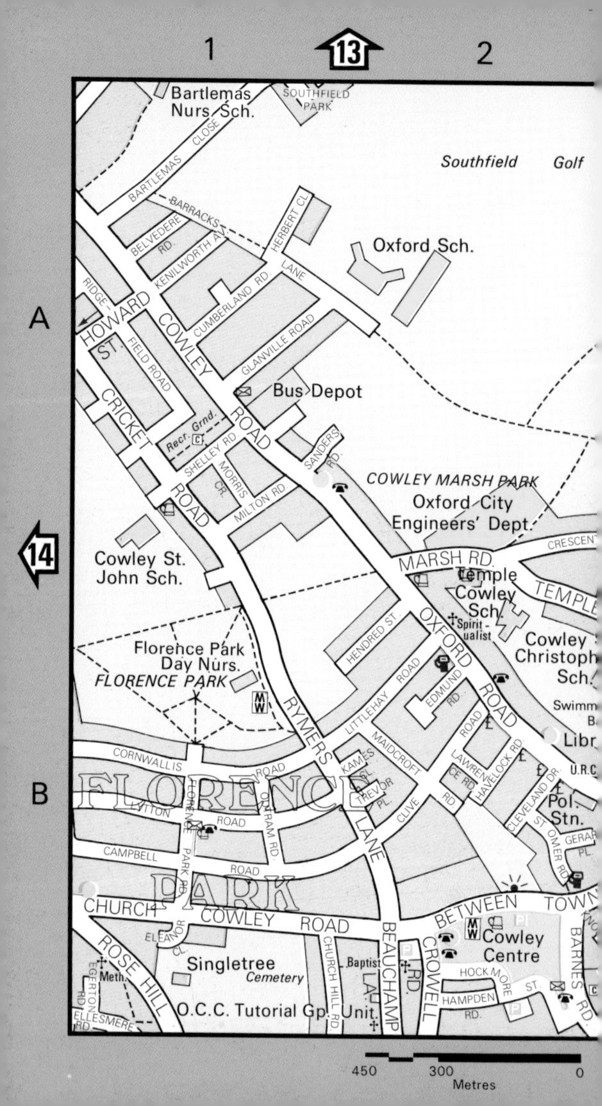
1
13
2
A
B
14
Bartlemas Nurs. Sch.
SOUTHFIELD PARK
Southfield Golf
BARTLEMAS CLOSE
BARRACKS
BELVEDERE RD.
KENILWORTH AV.
HERBERT CL.
LANE
Oxford Sch.
CUMBERLAND RD.
RIDGE-
HOWARD ST.
FIELD ROAD
COWLEY ROAD
GLANVILLE ROAD
Bus Depot
CRICKET ROAD
Recr. Grnd.
SHELLEY RD.
MORRIS CR.
MILTON RD.
SANDERS RD.
COWLEY MARSH PARK
Oxford City Engineers' Dept.
CRESCENT
MARSH RD.
Temple Cowley Sch.
TEMPLE
Cowley St. John Sch.
Spiritualist
Cowley St. Christoph Sch.
HENDRED ST.
OXFORD ROAD
Florence Park Day Nurs.
FLORENCE PARK
LITTLEHAY ROAD
EDMUND RD.
Swimm B.
Libr.
RYMERS LANE
MAIDCROFT
LAWRENCE RD.
HAVELOCK RD.
U.R.C.
CORNWALLIS ROAD
KAMES CL.
TREVOR PL.
CLEVELAND DR.
Pol. Stn.
FLORENCE PARK
LYTTON ROAD
FLORENCE PARK RD.
OUTRAM RD.
CLIVE RD.
ST. OMER RD.
GERARD PL.
CAMPBELL ROAD
CHURCH COWLEY ROAD
BETWEEN TOWNS
ELEANOR CL.
Cowley Centre
ROSE HILL
Singletree Cemetery
Baptist
CHURCH HILL RD.
BEAUCHAMP LA.
CROWELL RD.
HOCKMORE ST.
BARNES RD.
Meth.
EGERTON RD.
HAMPDEN RD.
O.C.C. Tutorial Gp. Unit.
ELLESMERE RD.
450
300
0
Metres

3
4
A
B
THE SLADE
BENSON RD.
DENE
BULAN ROAD
GLEBE
FAIR VIEW
LYE VALLEY
COVERLEY ROAD
LANDS
INOTT FURZE
TOWN FURZE
Florence Park Day Nurs.
CORN AV
PEPPER
LEIDEN ROAD
BROAD OAK
3 FIELDS RD
STUBBS AV.
PICKETT AV
TAVR Centre
MASCALL AV.
HORSPATH DRIFTWAY
Slade Hosp.
Fire Stn.
Telecom
KENNEDY CL.
JAMES WOLFE RD.
WAY
No.3 Group ROC
HUNTER CL.
YEAT'S CL.
FLETCHER RD.
CRANMER ROAD
FAIRFAX RD
RUPERT RD
RIDLEY RD.
BURTON PL.
Adult Training Centre
BRASENOSE DRIFTWAY
Recr. Grnd.
CORUNNA CR.
CRAUFURD RD.
BARRACKS LANE
TEMPLE
TURNER CL.
CRESCENT CL.
HORSPATH ROAD
Recr. Grnd.
St. Francis Sch.
Salesian Coll. (R.C.)
JUNCTION RD
HOLLOW
MARSHALL ROAD
SUNNYSIDE
BLEACHE PL.
WILKINS
WHITE ROAD
ROAD
PAGET ROAD
FAN-SHAWE PL.
NORMANDY CR.
St. John Bosco Sch.(R.C.)
COWLEY
SILKDALE CL.
OLIVER ROAD
FERN HILL ROAD
BURBUSH RD.
Our Lady's Sch. (R.C.)
GARSINGTON ROAD
AUSTIN ROVER GROUP LTD.
RING ROAD
COLERIDGE CL.
ST. LUKE'S ROAD
NAPIER ROAD
PHIPPS ROAD
BAILEY ROAD
COWLEY
Assembly
Plant
0
300
450
Yards

Notes on the maps

Map 1

- Golf Course
- Hospital
- Landmark/Place of Interest
- Park and Ride Terminus/Car Park
- Petrol station—selected
- View Point

Maps 2 – 15

Bureaux de Change—usually banks, but the one at the B.R. Station is open seven days a week, £

Car parks—a multi storey car park is denoted on maps 2-10 by a stack of symbols—the number denoting the number of floors,

Note; The car parks around the John Radcliffe Hospital complex are for the use of hospital staff and hospital visitors only.

Cycle Lanes—along arterial roads are also **Bus Lanes** and thus, within the hours shown on the signs are not to be used by private cars, and motor-cycles. Others may allow cycles only, depending on circumstances. Bus Lanes are devices to separate private from public traffic and allow buses to move more freely at peak times, They may also be used by taxis and emergency vehicles.

Footpaths represented on the map do not imply the existence of any public right of way.

Night Clubs/Discotheques—selected, ☆

Pedestrian street—such streets in Oxford are also open to buses, emergency service vehicles, and taxis.

Petrol stations—selected—opening times vary. Servicing may not be available at a petrol station,

All-night petrol station—TraveLodge Motel, Peartree Roundabout 1B2.

Shopping streets—areas where shops are most common are shown with a blue road casing.

Telephones—selected call boxes,

Toilets—where either sex is catered for alone, the symbol is amended, M men, W women.

Wine Bars—selected,

Gazetteer

Cross-references in the gazetteer refer only to the gazetteer

A

B

C

D

E

I

J

Jowett Walk. 7A1
Junction Road. 15B3
Juxon Street. 2A3

K

Kames Close. 15B2
Keble College. 3B3
Keble Road. 3B2
Keble Triangle. 3A2, 3B2
Kenilworth Avenue. 15A1
Kennedy Close. 15A4
Kennington. 1D2
Kettell Hall—*see Trinity Coll.*
Kidlington. 1A2
Kineton Road. 14A1
Kingdom Hall—Jehovah's Witnesses 13B1
King Edward Street. 6B4
King's Arms (K.A.). 6A4
Kings Lake. 1B1
Kingston Court. 2A3
Kingston Road. 2A3, 11B1
King Street. 2B4
Kitchen—*see Magdalen Coll.*
Knolles Road. 15B2
Kybald Street. 7B1

L

Lady Margaret Hall (L.M.H.). 11B3
Lady Spencer-Churchill College. (OxPoly) 1C4
Lake Street. 14A1
Language Teaching Centre 3B1
Larkrise First School. 14B4
Lathbury Road. 11A2
Latimer Grange. 13A4
Latimer Road. 13A4
Latner Building—*see St. Peter's Coll.*
Laud's Library—*see St. John's Coll.*
Law Library—*see St. Cross Building.*
Lawrence Road. 15B2
Leafield Road. 15A3
Leckford Place. 2A4, 11B1
Leckford Road. 11B1
Leiden Road. 15A4
Leopold Street. 14A4
Lewell Avenue. 12B1
Leys Place. 14A4
Library, City. 6B2
Library, Marston. 12A1
Library, Summertown. 11A1
Library, Temple Cowley. 15B2
Liddon Quadrangle—*see Keble Coll.*
Lime Walk. 13A4
Linacre College. 4B2
Lincoln College. 6B4
Lincoln Road. 14B1
Lindemann Building—*see Clarendon Laboratory.*
Link, The. 12B1
Linton House—*see St. Peter's Coll.*
Linton Lodge Hotel. 11A2
Linton Road. 11B2
Little Acreage. 12A2
Little Brewery Street. 13B1
Little Clarendon Street. 3B1
Littlegate Street. 9A2
Littlehay Road. 15B2
Littlemore. 1D3
Littlemore Hospital. 1D3
Littlewood's. 6B2
Lodge Way. 12A1
Logic Lane. 7B1
London Place. 13B1
London Road. 13A4
Long Bridges-Bathing Place. 14A2
Longwall Street. 7B2
Longworth Road. 2A2, 11B1
Lorry & Coach Park. 8A4
Love Lane. 3B4, 6A4
Lower Fisher Row. 5B4
Luther Street. 9B3
Lye Valley. 15A3
Lynn Close. 12B2
Lytton Road. 15B1

M

Macclesfield House—*see Oxfordshire County Council.*
Macmillan Building—*see Pembroke Coll.*
Magdalen Bridge. 10A3
Magdalen Coll. 7B3
Magdalen Coll. School. 10A4
Magdalen Deer Park—*see Magdalen Coll.*
Magdalen Grove—*see Magdalen Coll.*
Magdalen Road. 14A4
Magdalen Street. 6A2
Magpie Lane. 6B4
Maidcroft Road. 15B2
Maison Française. 11B2
Maitland Building—*see Somerville Coll.*
Maltfield Road. 12A3
Management Studies, Centre for —*see Templeton Coll.*
Manchester College. 7A1
Manor House. 12B4
Manor Place. 7A3
Manor Road. 7A3
Manor Road Ground. 13A4
Mansfield College. 4B1

N

O

P

Q

R

S

T

U

V

W

Y

Z

List of Abbreviations.

AA	Automobile Association.
ABC	Associated British Cinemas.
Auth.	Authority.
Bapt.	Baptist.
BBC	British Broadcasting Company.
Bldg/s.	Building/s.
BNC	Brasenose College.
BR	British Rail.
CAB	Citizen's Advice Bureau.
C&A	Clement and Auguste (Dept. Store).
Cath.	Cathedral.
CFE	College of Further Education.
Ch.	Church.
Cl.	Close.
Coll.	College.
Cons.	Conservancy.
Corpus	Corpus Christi College.
Cresc.	Crescent.
C.T.V.	Central Television.
Dept.	Department.
DoE	Department of the Environment.
D of T	Department of Transport.
DHSS	Department of Health and Social Security.
E	East.
Exp.	Experimental.
Ext.	Extension.
Gr.Orth.	Greek Orthodox.
HM	Her Majesty's (Government).
Hosp.	Hospital.
Hse.	House.
Inst.	Institute.
Isis	River Thames.
J.R.I	John Radcliffe Hospital, Phase I (Maternity).
J.R.II	John Radcliffe Hospital, Phase II (General).
kHz	Kilohertz.
km	Kilometre.
La.	Lane.
Lab.	Laboratory.
Lib.	Library.
LMH	Lady Margaret Hall.
Ltd.	Limited (Company).
m	Metre.
M&S	Marks and Spencer Ltd.
Meml.	Memorial.
Meth.	Methodist (Wesleyan).
ml.	Mile.
Mus.	Museum.
N	North.
NBC	National Bus Company Ltd.
NHS	National Health Service.

OCFE	Oxford College of Further Education.
OUP	Oxford University Press.
Oxfam	Oxford Committee for Famine Relief.
Oxon	Oxfordshire.
Oxpoly	Oxford Polytechnic.
Pav.	Pavilion.
Quad.	Quadrangle.
RAC	Royal Automobile Club.
RC	Roman Catholic.
Rd.	Road.
ROC	Royal Observer Corps.
S	South.
SA	Salvation Army.
St. Cat's.	St. Catherine's College.
Sch.	School.
Servs.	Services.
Soc.	Social.
Socy.	Society.
Sq.	Square.
St.	Saint.
St.	Street.
Stds.	Standards.
Stn.	Station.
TAVR	Territorial Army Volunteer Reserve.
Teddy Hall	St. Edmund Hall.
Terr.	Terrace.
The Broad	Broad Street.
The High	High Street.
The House	Christ Church (College)—Aedes Christi.
The Turl	Turl Street.
TS	Training Ship.
TV	Televison.
TWA	Thames Water Authority.
Univ.	University (College).
URC	United Reform Church.
VAT	Value Added Tax.
W	West.
Wk.	Walk.
yds.	Yards.